IMAGES of America
MOXIE

Dear Wilburn —

Thank you so much for the copy of your fantastic book which you so graciously donated to the Matthews Museum! Believe it or not, the tune of "Just Make it Moxie for Mine" is heard from time to time in our display room, and usually greatly appreciated.

Please accept this book as a token of our appreciation for the donation of yours —

[signature]
Co-Author

Moxie Nerve Food Company was an established leader in its time for using the power of advertising to project its desired image and sell product. Here, in front of the Boston Post Office, is a typical 1890s scene. A uniformed attendant dispenses ice-cold 5¢ glasses of delicious and healthful Moxie beverage from the rear of one of the company's many horse-drawn bottle wagons. (Courtesy of the Matthews Museum.)

ON THE COVER: Delivery wagons are parked in front of the Moxie Nerve Food Company headquarters at 68 Haverhill Street in Boston, Massachusetts, around 1890s. (Courtesy of the Matthews Museum.)

MOXIE

Dennis Sasseville and Merrill Lewis

Copyright © 2019 by Dennis Sasseville and Merrill Lewis
ISBN 978-1-4671-1656-5

Published by Arcadia Publishing
Charleston, South Carolina

Printed in the United States of America

Library of Congress Control Number: 2015955487

For all general information, please contact Arcadia Publishing:
Telephone 843-853-2070
Fax 843-853-0044
E-mail sales@arcadiapublishing.com
For customer service and orders:
Toll-Free 1-888-313-2665

Visit us on the Internet at www.arcadiapublishing.com

This could be a photograph of the elusive Lt. Clyde Ambrose Moxie, the discoverer of the key ingredient to the original Moxie Nerve Food. (Courtesy of the Matthews Museum.)

CONTENTS

Acknowledgments		6
Introduction		7
1.	The Moxie Mystique Is Born	9
2.	From Remedy to Refreshment	29
3.	Moxie on Top	41
4.	The Marketing Machine	69
5.	Struggles and New Owners	105
6.	A Distinctly Different Following	115
Bibliography		126
Index		127

Acknowledgments

The authors are deeply indebted to the Moxie Beverage Company, former owners of the brand, and their general manager Justin Conroy for granting us permission to use material covered by their trademarks. The Matthews Museum of Maine Heritage of Union, Maine, through its president, George Gross, granted valuable access to photographs and corporate documents in their extensive collection.

We are also grateful for the materials, including photographs, made available to us by the Maine Historical Society, the Jamaica Plain Historical Society (Massachusetts), the J.A. Tarbell Public Library of Lyndeborough, and the Manchester Historic Association (New Hampshire).

Several members of the New England Moxie Congress contributed information and/or photographs for this book, including Ira and Ann Seskin, John Longo, Dan DiBlasio, Jim Bennette, Dan Hovey, Jim Jansson, John Leheney, and Dennis Bruso. To them, we are highly grateful.

We also acknowledge the seminal works of Moxie lore created by authors Frank N. Potter (*The Book of Moxie* and *The Moxie Mystique*) and Q. David Bowers (*The Moxie Encyclopedia*), published at the time of Moxie's Centennial in the 1980s.

Erin Palazzo and Kathryn Sasseville greatly supported our efforts through volunteering their tough-love editing services on the manuscript. We are extremely appreciative of the guidance and patience provided by senior title manager Caitrin Cunningham and the staff at Arcadia Publishing.

Dennis Sasseville's love of Moxie and appreciation for Moxie culture can largely be traced to a former teacher, Henry Hicks of Needham High School in Massachusetts and past member of the Needham Historical Society: "Maybe they even drink Moxie in the Maldives Islands!"

Merrill Lewis, an off-and-on Moxie drinker since high school, got seriously "hooked" on Moxie and the New England Moxie Congress after buying his present house on Pine Island Pond in 1999. Merrill found that he was a half-block away from the iconic Moxie Bottle House, which was in the process of being dismantled, exposing its 1910-era label for the first time in 80 years.

Lastly, this book acknowledges the efforts of four Franks: Archer, Armstrong, Anicetti, and Potter. Quite frankly, without them and their contributions, Moxie might now be just a faded memory.

Unless otherwise noted, all images appear either courtesy of the Matthews Museum of Maine Heritage of Union, Maine, or the authors.

INTRODUCTION

On a fair day in May 1884, Lt. Clyde Ambrose Moxie stopped in to see his former Civil War comrade Dr. Augustin Thompson at Thompson's medical office located at 139 Market Street in Lowell, Massachusetts. He regaled his old friend with wild tales of his recent travels in the jungles of South America. There, Clyde related excitedly, he came upon a band of hearty natives who partook of a unique starchy root concoction for power and vigor whenever strenuous activity was at hand and the bolstering of the nerves was required. Clyde left his old friend with a generous sample of this newly discovered root substance. The good doctor promptly experimented with the South American discovery and ultimately incorporated it into a carbonated elixir that he called "Moxie Nerve Food" in honor of his old comrade.

Could that dashing Union officer pictured on page 4 be the Lieutenant Moxie of yore? This story was repeated over and over in the early years of Moxie Nerve Food, appearing on bottle labels and in widespread advertising. However, the story is completely fictitious. The surname of the handsome gentleman depicted in the photograph is lost to history, but it most certainly was not Moxie.

The fact that stories such as the Lieutenant Moxie yarn were created in the first place is emblematic of the era of remedies, nostrums, elixirs, pills, and powders, all to cure what ails one. This is especially true of the years between the end of the Civil War (1865) and the first decade of the 20th century. These were years of rapid change in America—years marked by westward expansion, a flourishing eastern economy, and a dazzling thrust of innovation and invention changing society and individual lifestyles everywhere.

Yet the common thread that almost all Americans faced, no matter their occupation or station in life, was the lack of widespread, competent health care. In the case of the central and western territories, trained physicians of any type were scarce. In the east, not every medical practitioner hanging out an office shingle was competent or necessarily affordable. Medical self-help practices largely ruled the day among a public that could be wary of professional healers but more than willing to embrace the often-wild claims of purveyors of commercial health products. The common advice of the day seemed to be, "No need to pay a physician, young man, when the pages of every city and small town newspaper are full of daily solutions to virtually any ailment that afflicts you, or you think might afflict you." Even if a person felt well for the moment, they could not risk being without the protection afforded through regular doses of Schenck's Seaweed Tonic, Hood's Sarsaparilla, or Lydia E. Pinkham's Vegetable Compound.

Into this consumer remedy stew of the 1880s' American landscape appeared yet another product—this one devoid of both alcohol and addictive drugs, unlike the vast majority of its competitors. The theme of the times seems to have been "your nerves are stressed and need help!" Dr. Augustin Thompson of Lowell had just the solution for the average citizen and knew that a body could not possibly be healthy without daily doses of his wonderful new product: Moxie Nerve Food. Thompson was no ordinary physician and possessed a distinct flair for advertising his wares.

This physician knew how to sell product. Thompson set the tone for the company's marketing orientation and style that carried through, in one form or another, for the next 130-plus years.

In its heyday, the Moxie Company's marketing gurus produced drink-dispensing bottle wagons, celebrity hand fans, "Horsemobiles," the iconic pointing Moxie Boy, attractive Moxie Girls, and signs, signs, signs everywhere. The company even experimented with huge wooden bottle-shaped refreshment stands. One of these stood for a time at the famous Coney Island amusement park before being moved to Pine Island Park in Manchester, New Hampshire, and was eventually converted to a private, lakeside summer cottage. That bottle is now the centerpiece of the Moxie Museum in Union, Maine, the birthplace of Dr. Thompson.

As successful as the Moxie brand was for its first half-century of existence, both the Great Depression and World War II took their toll on company sales and profitability. Competition for America's taste buds only increased in postwar years, and many soft drink brands flourished as the baby boomer generation prospered. In contrast, these were years when Moxie frequently struggled to find its way under changes in management and ownership.

The beverage Moxie, America's oldest continuously bottled soda pop (or "tonic," as many diehard New Englanders choose to call soft drinks) has generated a unique legacy of fact and fiction that continues to this day. At the beginning of the 20th century, the word "moxie" evolved to become a common term that implies vigor, stamina, nerve, and guts. But few realize that the term was derived from the drink itself, not the other way around. The drink Moxie came before the definition of moxie. This bitter-sweet cure-all elixir-turned-soft drink once outsold Coca-Cola but is not all that well known outside of its New England home turf today. The unique-tasting beverage indeed survives and has a dedicated following that keeps it from vanishing from store shelves.

This book chronicles the ups and downs, the twists and turns, and the mystique of this "Distinctively Different" but greatly misunderstood beverage. It tells of movie stars, presidents, and baseball greats who promoted this product and of a state that made Moxie its official soft drink. It tells of a huge annual festival held in a small town claiming to be the Moxie capital of the universe and of a passionate group of aficionados known as the New England Moxie Congress who are dedicated to the continued availability of this iconic product and the preservation of its legacy.

We sincerely hope that readers will enjoy this trip through Moxie Land and come to fully appreciate its patriotic wartime slogan, "What this country needs is plenty of Moxie!"

One

THE MOXIE MYSTIQUE IS BORN

Augustin Thompson was born November 25, 1835, in Union, Maine. He attended the local public school, where he developed a passion for reading and learning. He assembled a modest library that focused on history and science but also included many classical works, such as Shakespeare, as well as Latin and German texts. Thompson's brief apprenticeships with local blacksmiths and masons were largely unsuccessful, as his mind and heart were elsewhere.

The small Maine village of Union is just west of the coastal towns of Rockland and Camden. The area was primarily agricultural with a population under 2,000 residents. As with many small New England communities, Union and South Union embraced industrial progress through their numerous water-powered mills. Factories produced cabinets and carriages, boots and leather goods, musical instruments, and coffins.

Augustin Thompson served the Union army during the Civil War. He was wounded in action and developed tuberculosis as a result of a rifle butt blow to the chest. After the war ended, he was awarded honorary promotions to the rank of major and lieutenant colonel in recognition of his distinguished service. Thompson poses here (first row, center) in full uniform with some of his fellow soldiers.

After his Civil War service, Thompson's interests in medicine drew him to Philadelphia's Hahnemann Homeopathia College, where he graduated head of his class. Homeopathy is a system of alternative medicine created in 1796 by Samuel Hahnemann based on his doctrine of "like cures like." Preparations are referred to as homeopathics or remedies and use animal, plant, mineral, and synthetic substances. They are commonly written with reference to their Latin or faux-Latin names; thus, many of the 19th-century preparations listed ingredients such as *arsenicum album* (arsenic oxide) and *natrum muriaticum* (common sodium chloride). Introduced to the United States in 1825, homeopathic medicine quickly drew the attention of Boston area physicians. At the height of its popularity, there were nine homeopathic colleges, dozens of hospitals, and about 6,000 doctors claiming this specialty in the United States. By the 1920s, homeopathy was widely labeled a pseudoscience, and Hahnemann Medical College eventually became part of the Drexel University College of Medicine. This image comes from the Hahnemann Medical College 1898 souvenir book.

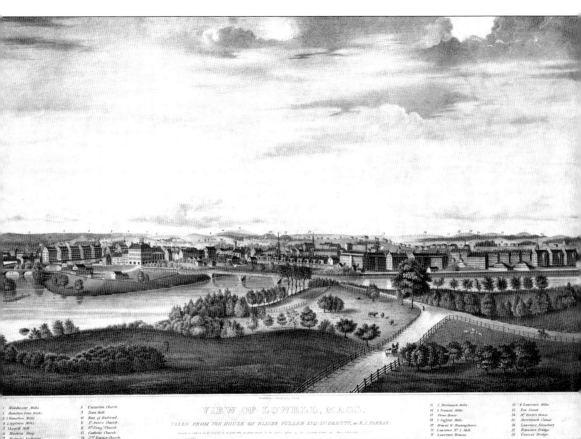

Lowell, Massachusetts, was already a thriving and important manufacturing center well before Augustin Thompson decided to call it home. Incorporated in 1826 as a planned manufacturing center, it became known as the cradle of the American Industrial Revolution due to the extensive series of textile mills that drew upon its abundant waterpower resources. By the 1850s, Lowell encompassed the largest industrial complex in the United States with 10,000 millworkers. It was here in the years after the end of the Civil War that both legitimate medical practitioners and purveyors of all types of home-style remedies thrived. J.C. Ayer Co. and C.I. Hood & Co. were among the largest and most popular of these suppliers. The map view of an already bustling 1834 Lowell provides an insight into its rapid early development and a glimpse into its promising future. (Courtesy of the Norman B. Leventhal Map Center, Boston Public Library.)

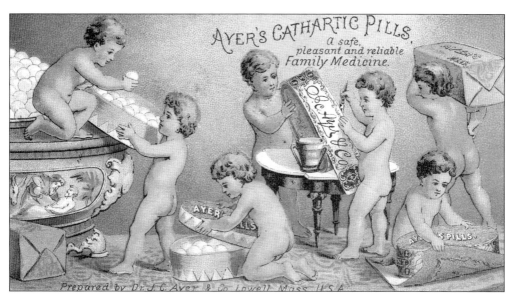

Thompson may well have been drawn to establish a medical practice in Lowell in 1867 because of the well-known Dr. J.C. Ayer, who sold remedies coast to coast from that thriving city. Spending up to $140,000 a year in advertising his packaged cures, Ayer never actually practiced medicine but was worth $20 million when he died in 1878. Ayer's Cathartic Pills were one of his most popular offerings.

In the years after the Civil War, remedies and patent medicines of all types flourished in the United States. They were dispensed right alongside phosphate sodas at the local druggist's counter or for purchase by the bottle either from those same dispensaries or from a host of other sources that were springing up at a dizzying rate.

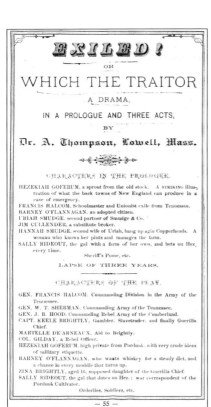

After Augustin Thompson settled in Lowell, he quickly developed a large medical practice, reportedly working 18 hours a day at one period. In spite of his work schedule, one of his other passions (drama) also came to the forefront. Thompson became a recognized playwright and had several dramas produced and performed on local stages.

By the mid-1880s, the genre of nerve foods and nerve tonics had become a well-established component of the fountain and bottled remedy trade. Although remedies of all types were already available to the public, Thompson's interests in medicine, and especially homeopathic remedies, led him to create his version for public consumption. He submitted his patent application for Moxie Nerve Food on September 8, 1885.

As a physician, Thompson purposed to develop a beverage that was devoid of the harmful substances contained in most remedies of the day, specifically cocaine and alcohol. Investing $15,000 of his personal funds into his commercial venture, he developed a formula composed of herbs and roots, stressing it could help cure alcoholism. Consistent with homeopathic theory, Thompson's personal belief was that diseases and ailments could best be treated gradually; administering remedies or cures in small, regular doses was part of his idealized design for Moxie Nerve Food. Moxie was first bottled and sold commercially in the early months of 1885, making it America's first bottled soft drink. It may well be, as some have claimed, that it was sold under this name in 1884 or that Thompson privately prescribed his remedy to his patients, perhaps as early as the 1870s. What is known for certain is that in 1885, Moxie caught on like virtual wildfire and Thompson focused his full attention on its production, advertising, and distribution.

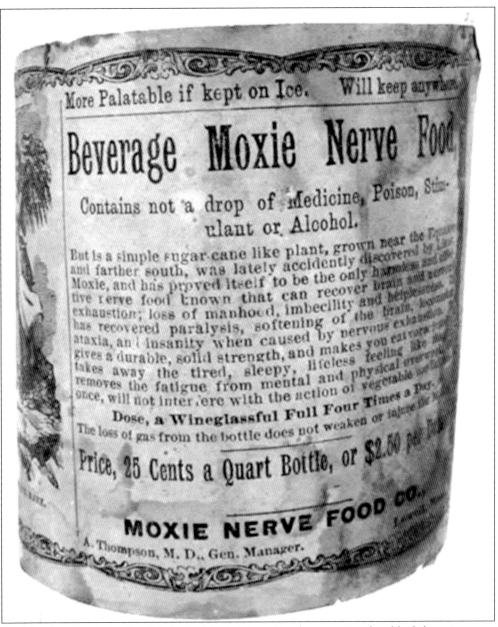

On March 1, 1885, the first labels for Moxie Nerve Food were printed as black lettering on a white background. Shipments to vendors consisted of both primarily quarts but also pints. Early production was individually wrapped in a protective paper covering, a practice that continued for the next 25 years. The paper labels on the bottles told the story as Thompson intended: "Contains not a drop of Medicine, Poison, Stimulant, or Alcohol," and "Dose, a Wineglassful Full [sic] Four Times a Day." He called it a food, not a medicine. The clear message: Moxie is good for you and should be enjoyed throughout each day. The label's admonition, "More Palatable if Kept on Ice," and the fact that Moxie Nerve Food was highly carbonated from the start may well speak to Thompson's attempt to counteract some of the beverage's inherent bitter taste from his so-called sugarcane-like plant.

PLANTES MÉDICINALES
1. Gentiane (Gentiana acaulis — Gentiana lutea)

PRODUITS LIEBIG : facilitent le travail culinaire.

Reproduction interdite   Explication au ver

In keeping fully with the mystique and mystery of key ingredients that most purveyors of 19th-century remedies and cures promoted, Thompson was initially quite guarded when it came to what made Moxie Nerve Food so special. His original claims of a sugarcane-like plant ingredient, or a turnip derivative, eventually gave way in future decades to proudly touting the beverage's true key ingredient: gentian root. This trade card's French text belies the fact that most gentian was imported to the United States from growing areas in France and Spain. No doubt the gentian in Moxie's formulation was imported in this fashion and not from "the equator and farther south," as claimed on the early bottle labels. As for the named accidental discoverer, Lieutenant Moxie, by all accounts, he appears to be a pure fabrication of Thompson's fertile mind for literary writings and his talent for developing creative advertising campaigns.

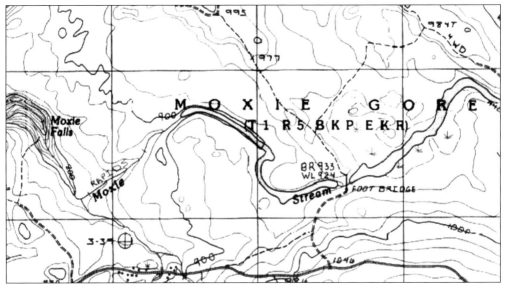

The most likely explanation of Augustin Thompson's use of the name "Moxie" for his new nerve food beverage involves looking no farther than his home state of Maine. Moxie Gore Township, Moxie Stream, Moxie Falls, and other geographical locales were recognized places in the Kennebec River Valley, as evidenced by published US Geological Survey topographic maps. On older maps, the name is sometimes expressed as "Moxy," or as in the case of this 1834 map of the Kennebec Valley Road to Quebec, Canada, the locale is presented as "Moxcy." In any case, it is clear that most residents of Maine in the 1800s would easily be familiar with these place names. (Above, courtesy of the US. Geological Survey; below, courtesy of the Old Canada Road Historical Society/ancestry.com.)

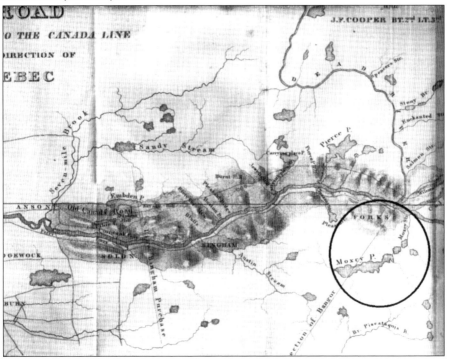

Moxie Mt. from Bald Mt. Station.

Moxie remains a well-known place name in the Kennebec River Valley with postcards of Moxie Mountain (above) and Moxie Falls (right) widely available. Nearby Moxie Pond and Moxie Stream are favorites of outdoorsmen and boaters. Reportedly, the name "Moxie" could have its origins from a Native American word for "dark waters." This explanation is actually quite logical, since many surface waters in Maine can be highly colored shades of brown or reddish brown from plant-derived, naturally dissolved organic compounds such as lignin and tannin. Gentian is a root extract, and the Moxie beverage has always been dark brown in color.

MOXIE FALLS

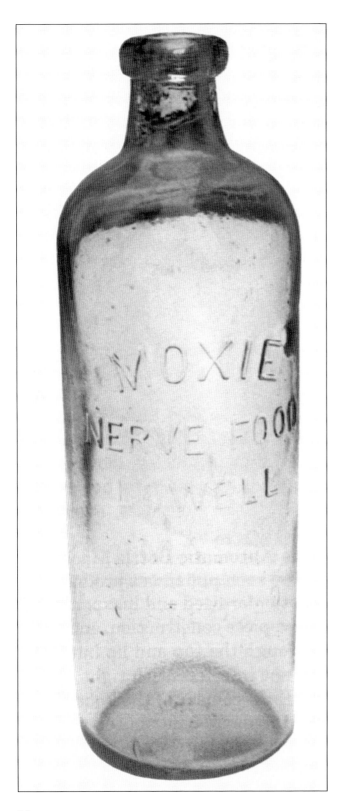

Most medicinal and remedy bottles of the post–Civil War years were square or rectangular glass of varying quality. For non-carbonated products on the market such as Steven's Sarsaparilla (Lowell) or Dr. Kilmer's Swamp Root Kidney, Liver and Bladder Cure (Binghamton, New York), such containers were perfectly suitable. But carbonated beverages require a stronger glass than was typically in use in the 1880s and a naturally strong shape—round. Thompson initially bottled Moxie Nerve Food in green champagne bottles since thick-wall round glass bottles were unavailable in any quantities. Collectors now call bottles in the illustration "blob-tops" because of the characteristic donut or blob opening, which was hand-finished at the glass factory.

The Moxie Nerve Food Company quickly contracted with the Lyndeborough (or Lyndeboro) Glass Company of southern New Hampshire for clear, round, embossed bottles to satisfy its rapidly growing product sales. Lyndeborough Glass Co. was a logical choice for Thompson since the factory had an established reputation for producing wood-molded glass bottles of superior blue-aqua clarity as well as strength. The glassworks also produced bottles for the well-known Hood's Sarsaparilla and Lydia E. Pinkham's Vegetable Compound, among others. The source of this high-quality, light-colored glass was a local deposit of quartzite (a quartz-rich granitic rock), which was mined for making into glass and firebrick, supplying materials to other glassworks as far away as Pennsylvania and New Jersey. Founded in 1866, the Lyndeborough Glass Co. burned in 1886 and closed for a time. This photograph of glassworkers and some of their products stored outside of the factory is undated. (Courtesy of the Lyndeborough [New Hampshire] Tarbell Library/Lyndeborough Historical Society.)

MOXIE
Nerve Food Company

OFFICE AND MANUFACTORY,
21 Branch St. Lowell, Mass.

A. THOMPSON, General Manager.
GEO. A. BYAM, Treasurer.
GEO. H. RICHARDSON, Clerk.

Presented by I. J. HUNGERFORD, Trav. Agt.

TRADE MARK.

MOXIE
PRICE LIST FOR 1887.

To Our Patrons:

Gentlemen:—Constantly appearing imitations of our goods, sold at the soda fountains, has caused us to substitute our XXXX and Syrup with a much stronger article called Triple Extract Moxie Syrup. This is three times as strong as the old syrup. While it gives us a very small margin, we hope dealers will use it instead of pushing us to the law for protection. They can make $46.50 above all, on a case of this at 5 cents a glass, which is better than can be done on a counterfeit.

PRICES.

X Moxie, $2.90 per case, $2.75 in gross lots.
XX Moxie, $4.00 per case, $3.80 in gross lots.
Triple Extract Moxie Syrup, $8.00 per case, or $3.50 per gallon.

Ten per cent. off to jobbers, freight free, who take five gross lots, two per cent. additional for cash. All goods sold on the rebate plan.

MOXIE NERVE FOOD CO.,
21 Branch Street,
Lowell, Mass.

Moxie was virtually a household name throughout the New England states by the close of 1885. The product also enjoyed growing recognition in other parts of the country as well. In 1886, a printed claim from Thompson stated that five million bottles of product had been sold in the first 14 months of operations. While this claim is dubious at best and exhibited Thompson's flair for passionately promoting his product, there is no denying that the good doctor had brought to market something close to "lightning in a bottle" in terms of grabbing the public's attention. Typically, distributors were sent concentrated syrup from which they bottled the Moxie product in their own operations with bottles procured locally, leading to some of the variants collectors enjoy today.

Moxie had to contend with both legitimate competitors as well as counterfeiters. Some say it was an era ripe for counterfeiting legitimate, successful products. Regulations to protect companies with a recognized "brand" were weak or nonexistent. Thompson counteracted imitators by preferring that his product be sold in bottles, not mixed from syrup at the soda fountain counter. Note the paper seal on the bottle top.

The transformation from apothecary to the American drugstore coincided with Moxie Nerve Food's meteoric rise in popularity. Fountain service in the new drugstores was quickly replacing the general store as the new locale for mixing commerce and socializing. The photograph shows a typical drugstore setting of the late 19th century, possibly in the Detroit area. (Courtesy of the Library of Congress.)

COCA-COLA
SYRUP ✣ AND ✣ EXTRACT.

For Soda Water and other Carbonated Beverages.

This "INTELLECTUAL BEVERAGE" and TEMPERANCE DRINK contains the valuable TONIC and NERVE STIMULANT properties of the Coca plant and Cola (or Kola) nuts, and makes not only a delicious, exhilarating, refreshing and invigorating Beverage, (dispensed from the soda water fountain or in other carbonated beverages), but a valuable Brain Tonic, and a cure for all nervous affections — SICK HEAD-ACHE, NEURALGIA, HYSTERIA, MELANCHOLY, &c.

The peculiar flavor of COCA-COLA delights every palate; it is dispensed from the soda fountain in same manner as any of the fruit syrups.

J. S. Pemberton,
Chemist,
Sole Proprietor, Atlanta, Ga.

One beverage that capitalized on the growing popularity of the soda fountain and would challenge Moxie for market share was Coca-Cola. The original formula was created by druggist John S. Pemberton in Columbus, Georgia. Like Thompson, Pemberton served in the Civil War and became addicted to morphine while recovering from a wound. In his search for a drug substitute, Pemberton developed and registered a tonic in 1885, French Wine Coca. At his Eagle Drug and Chemical Co., he used his fountain service to test flavors and concoctions on the public. After Atlanta passed prohibition legislation, Pemberton developed his nonalcoholic Coca-Cola formula. The first sales were at Atlanta's Jacob's Pharmacy on May 8, 1886 (more than a full year after Moxie's launch). Coca-Cola was originally sold for 5¢ a glass as a patent medicine with many curative properties as the early advertisement indicates. However, the first bottling of the beverage would not come until 1891 in Vicksburg, Mississippi. While ultimately Coca-Cola would grow to outpace Moxie in sales volume, Moxie is undisputed as America's first bottled carbonated soft drink.

Perhaps taking his lead from diversified Lowell competitors like the J.C. Ayer Company, Thompson sought to provide the public with a line of products, all related in one way or another to health and well-being. Sunbeam was a beverage designated as a remedy for alcohol consumption. Moxie Lozenges and J.S.Q. Nerve Food were essentially Moxie packaged in a solid and a powdered form.

PRICE LIST OF PREPARATIONS.

DR. A. THOMPSON, Lowell, Mass

FAMILY SAFEGUARD.
Per package .... .25
Per dozen .... $ 3.00
Per gross, delivered 36.00

SUNBEAM.
Per package .... .25
Per dozen .... $ 3.00
Per gross, delivered 36.00

MOXIE LOZENGES.
Per package .... .10
Per dozen .... $ 1.20
Per gross, delivered 14.40

J. S. Q. NERVE FOOD.
Per package .... .60
Per dozen .... $ 7.20
Per gross, delivered 86.40

30 per cent. off, to trade, on all of them. Bills 60 days.

General Office, 36 Varney St., Lowell, Mass

A. THOMPSON, M. D., Managing Agt.
LOWELL, MASSACHUSETTS.

BRANCHES ...
469 West Broadway, N. Y. City.
68 Beverly Street, Boston, Mass.
1002 N. Broadway, St. Louis, Mo.

One of the products offered by the Moxie Nerve Food Company in the early 1890s was Catarrh Cure, also called Family Safeguard. Its instructions stated, "Melt five pellets ... on your tongue before retiring and it will do away with the effects from exposure to colds and epidemic diseases during the day." The Cure was sold in unique shaped bottles.

Thompson continually demonstrated his talents for reaching the public marketplace with clever and attractive advertising approaches. In the coming years and decades, Moxie Nerve Food often utilized eye-appealing Moxie Girls to garner the public's attention to their product. Who could possibly resist a smiling lass declaring that the beverage was both delicious and strengthening? Sometimes, the dual message of "Delicious, Feeds the Nerves" would be associated with a Moxie Girl image. In later years, the Moxie Girls would often be represented by a recognized celebrity of those times, perhaps gracing a cardboard hand fan or a metal tip and change tray. Cardboard cutouts were typically of high quality and often lithographed in Germany for the Moxie marketing campaigns.

The summer of 1886 also saw the advent of the horse-drawn Moxie bottle wagon, soon to become a ubiquitous part of the Moxie marketing efforts and remain so up through the 1920s. The image of the Moxie Goddess and her sickle on the wagon's oversized bottle symbolized the harvest of Mother Nature's best for the inclusion in Moxie products. The American flag to the left of the Moxie Goddess image was a common patriotic theme for the beverage company. Note that the Moxie quart bottles (left wagon) were all individually wrapped in paper as another way of distinguishing the product from its competitors. Two of Augustin Thompson's sons appear in the photograph. Harry A. Thompson (future secretary-treasurer of the company) is the short lad in front of the wagon wheel, and Frank E. Thompson (future company president) is seated to the right on the bottle wagon. The others in the photograph are unidentified. (Courtesy of the late Virginia McElwee of Union, Maine, granddaughter of Augustin Thompson, via the Matthews Museum.)

Moxie certainly proved to be a commercial success, even outselling other popular beverages and remedies of its day. In 1889, however, Augustin Thompson decided to turn his primary attention back to his medical practice and, in later years, to the development of the Thompson Vitalizer, a device for infusing oxygen into the body for the benefit of one's nerves. He entered into an agreement with William Taylor, a Moxie sales agent from upstate New York, to lease the Moxie operations in Massachusetts. In return, Thompson received an annual fee of $5,000. In the late 19th century, this amount of income was sufficient for Thompson to live comfortably and pursue his other interests, including playwriting and publishing. He continued to be a presence in the production and sales of his creation but never again gave it his undivided attention. Augustin Thompson passed away on June 8, 1903, and is buried in the Lowell Cemetery with other local notables, including competitor J.C. Ayer.

Two

FROM REMEDY TO REFRESHMENT

As the end of the 19th century approached, Moxie Nerve Food sales continued strong, thanks to clever, constant advertising. Moxie was not the first consumer product to realize the power of advertising, but the company absorbed lessons from competitors and then went a step beyond. Horse-drawn bottle wagons with agents dispensing the healthful, refreshing beverage allowed Moxie high visibility throughout public spaces, especially in the Northeastern states.

Not everything the Moxie Nerve Food operation tried was successful, however. Some corporate initiatives were either ill-conceived or, more commonly, poorly implemented. Moxie failed in its attempt to establish a sustained marketing presence in the western states and territories, for instance. In contrast, other beverage and remedy concerns of the day, including many of Moxie's Lowell-based competitors, successfully marketed their products nationwide. The Moxie Company then resolved to strengthen and expand its solidly loyal base throughout the northeast from its Boston and New York centers. The 1896 hiring of Frank Morton Archer, a physician's son, proved to be critical to the company's future success. Archer had a flair for showmanship and advertising, two key ingredients that the company sorely required with the diminished involvement of founder Augustin Thompson. Archer oversaw most all of Moxie's successful sales campaigns.

Through its widespread advertising, Moxie positioned itself well as an essential ingredient in everyday life. In an era when the public expressed growing concerns over the farfetched claims of patent medicines and nostrums, Moxie advertised it was free of "injurious" ingredients. In the advertisement above, the woman informs that according to her doctor, Moxie is a "genuine nerve food and it is good for me because it contains no alcohol, narcotics, poisons, drugs or chemical preservatives." Pres. Teddy Roosevelt (below) never formally endorsed Moxie Nerve Food, but that did not stop the company from utilizing his healthy, sportsman's image to reassure consumers that its beverage supports the "Strenuous Life." Other advertisements stated that alcohol, tobacco, coffee, and other nerve stimulants "tear down the nerve tissue, while Moxie builds them up and supports the strenuous life."

The emphasis on nerves and nervous exhaustion was an extremely widespread concern in the second half of the 19th century, judging by the pervasive advertising evidence alone. In the June 6, 1895, copy of the *Youth's Companion* magazine, readers are reminded that this "Famous Nerve Food" is prescribed by physicians everywhere, sold by both grocers and druggists, and can counteract "Nervousness or Heat of Summer" exhaustion.

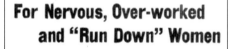

Naturally, Moxie was not alone in orienting its advertising toward American's worry over nerves and exhaustion. The Rexall Store produced and sold Americanitis Elixir "for Nervous, Over-worked, and 'Run Down' Women." Rexall pitched that its elixir was specifically designed for the exhausted nervous conditions wrought by the rush and tension under which Americas lived. People were informed that their nerves needed to be "toned and quieted."

Competition for consumers' attention was widespread and even came from Augustin Thompson's home state of Maine. Based in the coastal town of Belfast, Dalton's Sarsaparilla and Nerve Tonic combined two remedies into one: a blood purifier and nerve food. While East Coast manufacturers enjoyed an early start in establishing a marketplace dominance, other regions did not forsake the opportunity afforded by the post–Civil War westward expansion. Patent medicines were a staple in the sparsely populated territories where physicians and hospitals were generally scarce. Valentine Hassmer's Lung & Cough Syrup of San Francisco was just such a product sought out by homesteaders, prospectors, and the professional class alike to serve their medical needs.

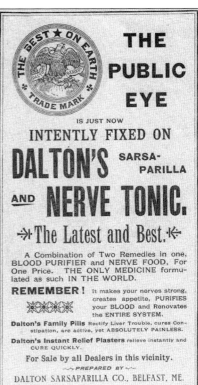

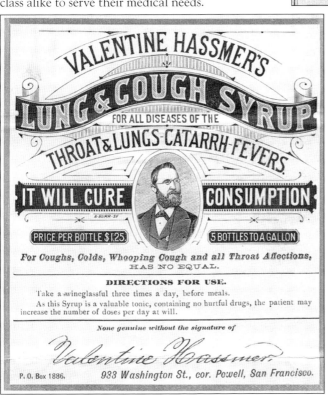

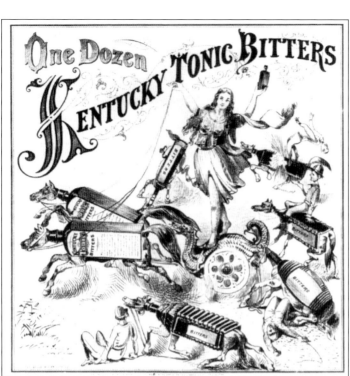

In the 19th century, a merchant's choices for advertising their products to the public was fairly limited to either print or the spoken word. Certainly word of mouth, physician recommendations, and written testimonials were popular and effective means to selling remedies and patient medicines of all types. Merchants of all types realized early on that images could be even more powerful than words. Handbills, posters, trade cards, and newspaper advertisements were common avenues for informing and enticing the public. Trade cards (above) and print advertisements (right) were both an imaginative and colorful means to draw attention to a merchant's products and were used extensively in promoting patent medicines. (Both, courtesy of the Library of Congress.)

Few remedies or nostrums of the era could rival the success of Lydia E. Pinkham and her Vegetable Compound. From yet another northern Massachusetts city, Lynn, Pinkham concocted her legendary herbal-alcoholic product and pitched it to women as a relief for menstrual and menopausal symptoms. Like Augustin Thompson who followed her, Pinkham had her pulse on the public's health concerns and oriented her extensive advertising campaigns accordingly. Pinkham's formula was sold as a remedy for "those painful complaints and weaknesses so common to our female population." That population enthusiastically responded to her product, and soon, she was grossing $300,000 per month. Pinkham's own motherly image became the iconic symbol for her products and was used in company advertising long after her 1883 passing. By 1900, hers was one of the most recognized female images in America.

INKHAM'S . PROVERBIAL . PHiLOSOPHY. . . . .

Coming events cast their shadows before.

The feeling of utter listlessness, lack of energy, desire to be alone, or the "don't care" feeling, are all shadows of coming events. No woman should permit those symptoms to gain ground, for, being forewarned, she should be forearmed. *Lydia E. Pinkham's Vegetable Compound* will disperse all those shadows. It goes to the very root of all female complaints, renews the waning vitality, and invigorates the entire system. Surely such letters as this will support our claims:

"Reach for a vegetable instead of a sweet"

Bitters represented one of the most popular categories of remedies as early as the 18th century. Noah Webster's 1849 dictionary defined it as "a liquor in which bitter herb roots are steeped; generally a spirituous liquor." Medicinal books of the era reported that bitters were thought to exert a "tonic power on the digestive organs." Brown's Iron Bitters was typical of the day, touting its overall quality as a "Valuable Family Medicine," in spite of its composition of almost 20 percent grain alcohol. Alarmingly, many bitters and other patent medicines on the market had twice Brown's alcoholic content, as documented by the state of Massachusetts. By the beginning of the 20th century, a tide of public and governmental concern was rising with respect to all patent medicines. The image below is from the June 1904 issue of the *California State Journal of Medicine*.

ALCOHOL IN "PATENT MEDICINES."

The following percentages of alcohol in the "patent medicines" named are given by the Massachusetts State Board Analyst in the published document No. 34:

Medicine	Per cent. of alcohol (by volume)
Lydia Pinkham's Vegetable Compound	20.6
Paine's Celery Compound	21
Dr. Williams's Vegetable Jaundice Bitters	18.5
Whiskol, "a non-intoxicating stimulant"	28.2
Colden's Liquid Beef Tonic, "recommended for treatment of alcohol habit,"	26.5
Ayer's Sarsaparilla	26.2
Thayer's Compound Extract of Sarsaparilla	21.5
Hood's Sarsaparilla	18.8
Allen's Sarsaparilla	13.5
Dana's Sarsaparilla	13.5
Brown's Sarsaparilla	13.5
Peruna	28.5
Vinol, Wine of Cod-Liver Oil	18.8
Dr. Peters's Kuriko	14
Carter's Physical Extract	22
Hooker's Wigwam Tonic	20.7
Hoofland's German Tonic	29.3
Howe's Arabian Tonic, "not a rum drink"	13.2
Jackson's Golden Seal Tonic	19.6
Mensman's Peptonized Beef Tonic	16.5
Parker's Tonic, "purely vegetable"	41.6
Schneck's Seaweed Tonic "entirely harmless"	19.5
Baxter's Mandrake Bitters	16.5
Boker's Stomach Bitters	42.6
Burdock Blood Bitters	25.2
Greene's Nervura	17.2
Hartshorn's Bitters	22.2
Hoofland's German Bitters, "entirely vegetable"	25.6
Hop Bitters	12
Hostetter's Stomach Bitters	44.3
Kaufman's Sulphur Bitters, "contains no alcohol" (as a matter of fact it contains 20.5 per cent of alcohol, and no sulphur)	20.5

Elixirs, nostrums, and remedies were sold under the pretense that they could cure virtually any ailment known to man. These so-called patent medicines of the 19th century were not actually patented at all since chemical and drug patents did not come into use in the United States until 1925. Further, submitting to a patent would mean having to disclose formulas and ingredients, something the promoters sought to obfuscate, if not totally avoid. Newspaper and magazines of the day were somewhat complicit in validating, or at least not openly questioning, products' wild health claims. This is perhaps not surprising, as print media derived substantial income from running advertisements for these unregulated products. Publisher Edward William Bok built a successful career focusing on women's issues, and as editor of the *Ladies' Home Journal* in 1892, he announced the magazine would no longer accept patent medicine advertising. This bold move helped begin to turn public sentiment and paved the way for governmental reforms of the patent medicine industry. (Courtesy of the Library of Congress.)

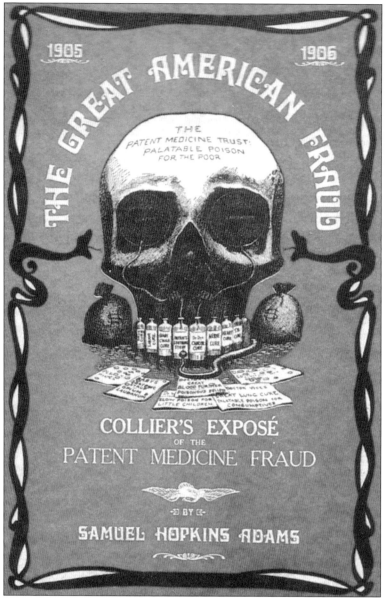

In the last quarter of the 19th century, journalists and other investigators began to document and publicize the dangers associated with many of the categories of patent medicines that were on the market. Alarmingly, drug addiction cases and even deaths that could be tied to patent medicines were far more common than the general public realized. In addition to products sold to the general population that were based on a high-alcohol content, many patent medicines contained additive narcotics such as opium or cocaine. Even patent medicines formulated with compounds that might have health and wellbeing benefits in limited doses (caffeine for example), often contained those substances at levels that risked toxicity to human life. In 1905, Samuel Hopkins Adams published an expose entitled "The Great American Fraud" in the magazine *Colliers Weekly*, attacking 264 remedies by name. The magazine's article represented an important public opinion tipping point that eventually culminated in the passage of the federal Pure Food and Drug Act of 1906.

In 1882, Harvey W. Wiley, a crusading pure foods chemist was appointed to lead the federal Division of Chemistry, later to be incorporated into the Food and Drug Administration. Wiley was passionate regarding the need for improved regulation of both the food and the drug industries. His research and leadership directly led to the passage of the seminal Pure Food and Drug Act. The 1906 statute did not create an outright ban on the alcohol, narcotics or questionable stimulants in patent medicines, but rather, it curbed some of the most misleading, overstated, or clearly fraudulent claims being made by the products' purveyors. The future for such products was made perfectly clear: either clean up the product formulations and claims or risk being shunned by the consumer and scrutinized by regulators. (Both, courtesy of the Library of Congress.)

The Food and Drugs Act June 30, 1906.

THE GENERAL PURPOSE AND SCOPE OF THE ACT.

1. THE PURPOSE OF THE ACT.

The purpose of the Act is indicated in the title,[1] to prevent so far as within the constitutional power of the Federal Government to do so, the manufacture, sale or transportation of food or drugs which are either so adulterated as to be below the standard of quality expected by the purchaser, or are so poisonous or deleterious in themselves or by reason of the addition to them of poisonous or deleterious coloring or preservatives as to be injurious to health, or which are misbranded or labeled in such a way as to deceive the public as to their character, quality, locality of origin, or manufacture.

MOXIE 22nd Year

The merit of MOXIE will be maintained during our twenty-second season — twenty-one years of increasing custom and popularity spell REAL MERIT — the perfect cleanliness of manufacture — the tonic quality — the delicious satisfying taste — will still be distinctive of MOXIE; will still make new friends for MOXIE and new customers for MOXIE dealers during the season of 1906.

While the passage of the 1906 Food and Drug Act spelled trouble for most patent medicine, Moxie Nerve Food found itself in a good position. Always marketed as a beverage rather than a hard medicine, the drink contained no alcohol or narcotics. Moxie's pureness was emphasized and touted as an essential in everyday life while competing beverages that contained recognized harmful ingredients were publicly chastised for their contents. Distribution to the West Coast had ceased by 1906, but eastern sales were booming, pumped up by the company's copious advertising. In that year, some 16,840 metal signs and 9,450 cardboard cutouts in the shape of a Moxie bottle were ordered and delivered to the company's advertising department for distribution. Moxie print advertisements celebrated 1906 as the beverage's 22nd year—an interesting claim, as that meant the beverage was birthed in 1884, not the patent date of 1885. Future advertising and bottle labels would wholeheartedly adopt the 1884 date.

Three

MOXIE ON TOP

Born in 1862 in Lincoln, Maine, Frank Morton Archer was the son of a physician, but he exhibited natural talents in sales and promotion. Joining the Moxie organization in 1896 as an office worker, Archer would rise through the ranks to steer the enterprise to its successful run in the early 20th century. Many of the company's advertising pieces would commonly display "Frank Archer says," almost as a trademark phrase.

ICE-CREAM SODA!

We have added in connection with our Soda Fountain, Ice Cream Soda, Sherbet,

Moxie Nerve Food,

and shall at all times take the lead in everything

Cool and Refreshing

Universal Verdict that our Soda is the
BEST IN TOWN.

Bananas,   Pineapples,   Oranges,
   Lemons,   Confectionery.

ICE - CREAM!

Even though the Moxie Company strongly preferred to sell soda fountain proprietors the highly carbonated Moxie Nerve Food in bottles and not the syrup, the beverage was extremely popular, and most establishments were eager to carry and promote the brand any way they could. In this c. 1920 scene, Moxie is prominently promoted, including a display of the pointing Moxie Boy on the back wall, beckoning customers to partake in the delights of Moxie. A popular Moxie Girl cardboard fan peers out from behind the woman's right elbow, and another Moxie poster graces the wall behind the glass drink dispenser. In a different establishment's signage, Moxie Nerve Food is given high visibility to all potential customers.

In the 18th century, chemists developed a way to saturate water with carbon dioxide, leading to a British patent in 1807. The carbonated water was typically referred to as soda water, even though it contained no soda (sodium bicarbonate). Soda fountains began in Europe but found their greatest success in the United States in pharmacies, ice-cream parlors, candy stores, five-and-dime stores, and train stations. The interiors of Collins Pharmacy on Long Island in 1915 (above) and the People's Drug Store in Washington, DC, in the 1930s (below), reflect the integration of the all-important soda fountain into the establishment's business activities. (Both, courtesy of the Library of Congress.)

In addition to their 27¢ per gallon Socony gasoline, travelers also required refreshment (and sometimes even lodging). This commercial roadside stand and inn (above), presumably located somewhere in New Hampshire in the 1920s, offered a variety of items from eggs to Moxie. The cutout display of the Moxie Boy sitting on a wooden crate rates a prominent position in the front of the stand and reminds customers what to order to feel refreshed. Another country store in an unknown location (below) seems to be messaging that Moxie has a season, presumably a tongue-in-cheek encouragement for patrons to stock up on the beverage before it is gone.

The neighborhood store was certainly Moxie's friend, especially since the company always preferred to sell its highly carbonated beverage cold and in factory-filled bottles rather than mixed at fountain services by some possibly untrained drugstore employee. Neighborhood stores were the lifeblood of most rural and urban areas alike, with families making daily treks there for necessities and treats and schoolchildren and tired workers stopping in on their way home in the afternoons. This little store in Vermont displays Moxie advertising in both storefront windows (Moxie Boy's head is obscured by another sign in the left window). Often family-run, such small convenient stores in the Boston area and other parts of Massachusetts were often called "spas," especially if they had, or once had, a soda fountain. The term "spa" referred to the health claims of the fizzy carbonated beverages offered inside. A few self-named spas can still be found in older city areas, primarily around Boston.

Moxie tirelessly spread its advertising by any means it could, thanks in large part to the persistent drive of Frank Archer. A large billboard graces a busy downtown Bangor, Maine, around the 1920s (above). One of Moxie's favorite promotions with stores was to offer them a "compensation case." This program varied somewhat over the years but generally consisted of giving the store proprietor one or more free cases of Moxie, with glasses or mugs sometimes included. The proprietor then agreed to "conspicuously display Moxie and Moxie signs in the windows of the store for at least two weeks." This neighborhood store (below) left little doubt as to which beverage held its loyalty.

Open-air refreshment stands carried Moxie along with other beverages (above). Moxie Girl appliqués are visible on the side windows while the stand's front sports flanking Moxie signs. But the center position was reserved for advertising Coca-Cola (in straight-sided bottles), certainly a foreshadowing of the dominance this brand would eventually register on the global marketplace. The Pine Tree Filling Station, a compact roadside eatery, store, and gas station (below), displayed a large oval Moxie sign on its roof. The store also displayed several Fro-joy Ice Cream signs. Fro-joy, later to become Sealtest, used the image and photographs of Babe Ruth extensively in the 1920s, especially in baseball card giveaways with each cone purchased. Some 30 years later, Moxie would also use a sports hero, Red Sox baseball legend Ted Williams, to promote both Moxie and Ted's Root Beer.

Moxie was virtually always present at New England's agricultural, state, and county fairs, frequently employing one or more Moxie bottle wagons to maximize sales. At the fair in Brockton, Massachusetts, a Moxie bottle wagon can be seen to the right on the midway. The wagons were created in 1886 but persisted into the 1920s, well after Moxie vehicles were the dominant mobile advertising device. Their popularity and use likely peaked around the turn of the century, approximately the same period as represented by this fair scene. One bottle wagon's 1900 itinerary included fairs in Lewiston, Old Orchard Beach, and Bangor, Maine; Rochester and Portsmouth, New Hampshire; and Taunton, Reading, and Brockton, Massachusetts. Bottle wagon inventories were always recorded according to cases sold, broken, stolen, and given away. It was customary to give fair officials and local constables a free drink of Moxie as goodwill gestures. By and large, stores and other stationary establishments did not appreciate the mobile bottle wagons, believing they took too much business away from them.

A c. late-1920s "food court" adjacent to a racetrack, location unknown, may also be a part of a local or state fair event. The food and beverage booths appear to be run by different vendors, each advertising their own offerings. Moxie is well represented among the various stands as is competitor Coca-Cola. Two other competing soft drink brands are also well represented, Whistle and Dr. Swett's Root Beer. Whistle Orange was a popular drink first created in 1925. Dr. Swett's was a storied root beer brand alongside of famed Hires. Both products predated the advent of Moxie but were originally sold in extract form for home brewing or to be dispensed at soda fountains—not commercially bottled. Dr. George Swett's Root Beer was part of a line of products from his Eclectic & Botanic Medicines business based in Boston. The long-lived root beer (1845–1959) was frequently advertised as a temperance drink with a tagline of "A drink that's good for you," not at all dissimilar to Moxie's persistent advertising claims.

Vermonter Calvin Coolidge was the country's vice president in 1923 when Pres. Warren G. Harding died suddenly. Coolidge was at the family home in rural Plymouth Notch, Vermont, which had neither electricity nor a telephone. When word of Harding's passing reached the home in the middle of the night, Coolidge dressed, said a prayer, and went down to the parlor where some reporters had gathered. He was immediately sworn into office by his father, a notary public and justice of the peace. Some accounts then have Coolidge returning to bed, but one account has him exclaiming, "We should celebrate this somber occasion." Coolidge, his wife, father, and the reporters crossed the street to Miss Cilley's Store, where the president ordered a round of Moxie for the gathering. Finishing his drink, Coolidge pulled out his change purse and extracted a solitary coin to pay for his beverage before leaving. This story may or may not be factual, but it is well-documented that "Silent Cal" was genuinely fond of Moxie. This photograph was taken in 1923. (Courtesy of the Library of Congress.)

The women's rights movement started in 1848 in Seneca Falls, New York, and grew from there, especially turning its focus to winning the right to vote. On March 3, 1913, seven years before passage of the 19th Amendment, a suffrage parade marched in Washington, DC (above). As part of this awakening, the temperance movement in general—and specifically, the Women's Temperance Union—was a powerful social force when Augustin Thompson introduced his Moxie Nerve Food to the world in 1885. Like Hires Root Beer and other nonalcoholic beverages of the day, Moxie frequently played up the temperance angle; in this case, with a humorous advertisement for its lozenges (right). (Above, courtesy of the George Grantham Bain Collection, Library of Congress.)

Temperance movements have had their successes. Prior to the Civil War, Maine and several other states enacted—and then within a few years rescinded—laws prohibiting the sale of alcohol. Federal Prohibition was tried between 1920 and 1933 before the law was repealed. This Bureau of Prohibition sign (left), being showcased by three somewhat dour-looking gentlemen, was used in a 1930s campaign to stop and search vehicles suspected in bootlegging activities. The 1927 photograph below, taken in Portland, Maine, displays a confiscated stash of illegal whiskey and cognac bottles; some transported using Moxie's sturdy wooden cases—an unauthorized use that could not have pleased the company. (Left, courtesy of the Library of Congress; below, Collections of Maine Historical Society, No. 14557.)

In 1925, Moxie's aging facility at Leight and Varick Streets in New York City was on its way out, slated for demolition to make room for one of the entrances to the Holland Tunnel, then under construction. At this point, most bottled Moxie was being shipped to distributors from the Boston operation at Haverhill Street, which itself was looking for a new home. Moxie's management found a solution in a Prohibition-idled brewery in the Roxbury–Jamaica Plain section of Boston, south of the center city. The 85,000-square-foot building was purchased for $180,300 in May 1925 and readied for a 1926 move-in. Frank Archer named the complex "Moxieland" and envisioned the location as more than just a bottling plant. Archer saw the new facility as an opportunity to establish a civic center of sorts, a focal point for the community of Boston, and advertised it as such.

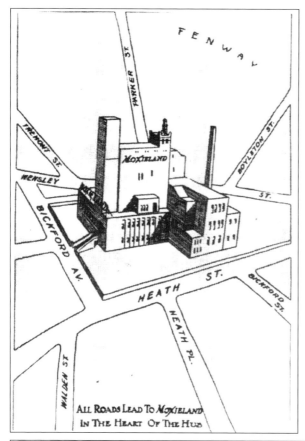

Bounded by Parker, Old Heath, Heath, and Bickford Streets, Moxieland was located in a hub, and Frank Archer presented it accordingly. Archer's published diagram (left) proudly proclaimed that "All Roads Lead To Moxieland In The Heart Of The Hub." The city of Boston itself had long been referred to as "the Hub," since settled satellite areas and major roads radiated from it like a wagon hub and spokes. Ever the marketer, Archer had the industrial roof of Moxieland painted with its name visible by air, especially from flights bound for the city's Jeffrey Field, later to become Logan International Airport (below).

In truth, there was "Moxie Land" before there was "Moxieland." At some point, Frank Archer began to refer to Boston's Haverhill Street bottling location as Moxie Land and encouraged visitors as a form of product promotion (right). In this 1922 advertisement from the *Radcliffe News*, Archer reaches out to New England college students and faculty inviting them to discover the source of the "delicious bittersweet flavor" of his "carefully compounded tonic beverage." Moxieland workers were always referred to as "associates" during this period, and all production staff wore white uniforms. The occasion for the posed group photograph (below) of May 8, 1931, is unknown.

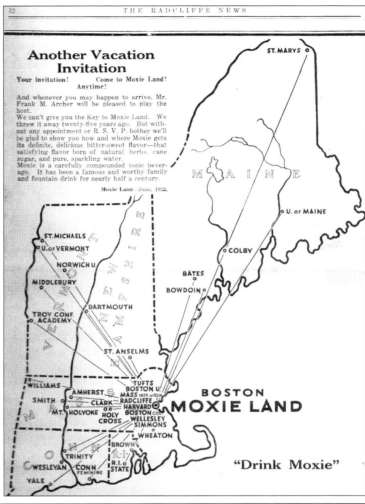

Moxie extract was compounded on the fifth floor of Moxieland and three 1,000-gallon tanks were used to feed the bottling line on a lower floor (left). About 5,200 cases of beverage per day were produced in three different size bottles, including the seven-ounce bottle introduced by Archer (below). The glass bottles were procured from the American Bottle Company in Toledo, Ohio, with three railcars per week arriving at Moxieland to supply the plant's robust production levels. The production facility itself was considered a model plant in its day with large windows for natural light and all interior surfaces painted white, except for red tile floors.

Temperance groups were especially pleased that Moxieland occupied a former brewery operation that had closed due to the national Prohibition laws. In furthering his civic-center concept, Archer had a penthouse room remodeled and opened it for group meetings and social functions and associated tours of the production facility and its clean and tidy work areas (right). The Women's Christian Temperance Union was especially fond of using Moxieland's repurposed space for their gatherings, relishing the symbolism of the building's former activities. Moxie management purchased the Pureoxia line of beverages in January 1931 and used the upper floors of Moxieland to house this locally popular line of bottled soft drinks (below).

PUREOXIA

DRY
GINGER ALE

THIS ginger ale is in **every respect** the equal of any in the world, and possesses in addition the **unique advantage** of being **made from distilled water.**

This permits of a **much more perfect blend,** than if spring waters were used in its manufacture, and consequently results in a much **finer flavor** and **greater brilliancy.**

Nothing but the purest ingredients are employed in its preparation, which makes it **especially desirable** for **children** and **invalids.**

A **special advantage** it possesses over the old style variety is that **corks** are replaced by **caps** as stoppers.

Bottles are white flint glass and hold 2 full glasses each. They are packed 24 in an upright partitioned case, where they may remain until used. No covers or excelsior used in packing.

PRICE.

Per case of 2 doz. pint bottles $2.45
Rebate on empty bottles and box 95

(Making net price 75c. per dozen.)

Also put up in half-pints and quarts.

Tel. { 1147 / 1542 } Back Bay.   Or Your Grocer.

The Pureoxia Company was formed in 1899 by Harry A. Edgerly, who formerly owned several Florida hotels where he learned and practiced the art of total customer satisfaction. In entering the beverage business, Edgerly desired to produce a superior product of the highest purity. With a $100,000 capital investment, he established his company at 100 Norway Street in downtown Boston. Only the finest ingredients and pure distilled water were used with state-of-the-art blending and bottling equipment designed to assure full hygienic conditions. By far, Pureoxia's flagship product was its sparkling superior blend of Golden Ginger Ale, generally sold in three bottle sizes. The company's sales covered all of New England, and one innovation was to use electric trucks to move its products, with swift automobiles used to make timely deliveries to parts of its territory. Pureoxia's trademark was successfully registered on January 1, 1907, and only expired in 1988.

In addition to its vaunted Golden Ginger Ale, Pureoxia produced a variety of beverages and flavors over its 50 years of production. Known flavors included Pale Dry Ginger Ale, club soda ("carbonic"), orange, lemon, birch beer, root beer, sarsaparilla, and Sparkling Grape (reportedly highly carbonated to the delight of youngsters). The club soda was sold in distinctive egg-shaped bottles (pictured here on the left). Pureoxia's paper-labeled flavored beverages such as this Aromatic Lemon Soda (right) stressed "Extra Quality," "High Grade," and the use of distilled water in its formulation.

Black Cherry Soda appears on no known listing of the various Pureoxia flavors, yet here is a paper-labeled capped quart bottle full of that very product.

59

Throughout the 1920s, Moxie chemists tried to develop a ginger ale product that could match Pureoxia's product in flavor and quality but never found a formula that satisfied management, especially Frank Archer. When Pureoxia founder Harry Edgerly looked to sell his storied brand, he looked only to the Moxie Company. Edgerly felt strongly that only Moxie represented the high standards his company embodied, coupled with the modern equipment and hygienic production practices he valued. On January 1, 1931, Pureoxia was acquired by the Moxie Company. Subsequently, both brands commonly shared print advertising and signage space (above). Through a loyal following, Pureoxia did well for Moxie throughout Boston and its suburbs. This monthly statement of January 1941 to the City Hall Spa (convenient store) in Cambridge, Massachusetts, shows two cases of Moxie sold to the establishment but six of Pureoxia (below).

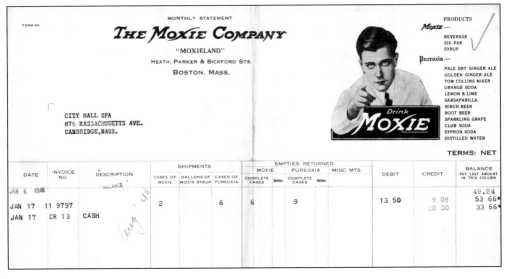

The superiority of Pureoxia's Golden Ginger Ale no doubt reflected both the quality of its ingredients as well as the formulation process. Moxie Nerve Food, of course, had its special ingredients and preparation methods as proudly pronounced in print advertising (right). The description indicated that while the Moxieland facility may be open to all, the formula itself was certainly not. The popular beverage's combination of unique taste and health benefits should be the public's primary interest. A page from the 1927 laboratory notebook of James Penny, longtime Moxie chemist, reveals the procedure for extracting gentian root using a Moxieland "Packing Percolator" (below). The procedure is much more reflective of straightforward c. 1920 food science rather than any Frank Archer secret alchemy.

MOXIE is inimitable, for its formula is secret. Many have attempted to imitate it and have tried to simulate the name, the bottle, the package and other distinctive features so well known and recognized as indicative of the genuine. But substitutors make little headway in the market even before we discover them. After we discover them they make no progress at all, as the numerous judgments, decrees and injunctions printed in this book well attest.

MOXIELAND is open to everybody; pure food experts, doctors, chemists, bottlers and other trained observers are especially welcome. The sanitary process of producing MOXIE is as essential as the formula itself and therefore all MOXIE extract is made solely at the MOXIELAND laboratory, under strictly hygienic conditions. The secret formula is a scientific compound of distinctive ingredients and the result is an absolutely safe and wholesome tonic beverage for all.

The system of producing MOXIE extract is as unique as MOXIE itself.

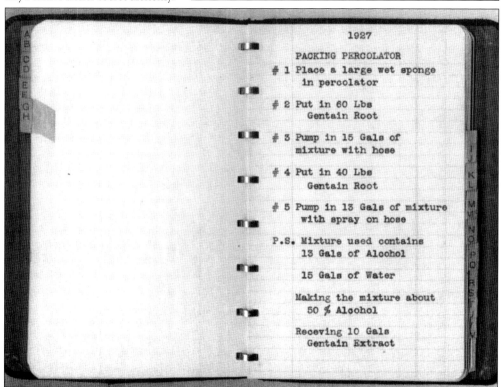

```
                    1927

            PACKING PERCOLATOR
      # 1 Place a large wet sponge
              in percolator

      # 2 Put in 60 Lbs
              Gentain Root

      # 3 Pump in 15 Gals of
              mixture with hose

      # 4 Put in 40 Lbs
              Gentain Root

      # 5 Pump in 13 Gals of mixture
              with spray on hose

      P.S. Mixture used contains
              13 Gals of Alcohol

              15 Gals of Water

            Making the mixture about
              50 % Alcohol

            Receving 10 Gals
              Gentain Extract
```

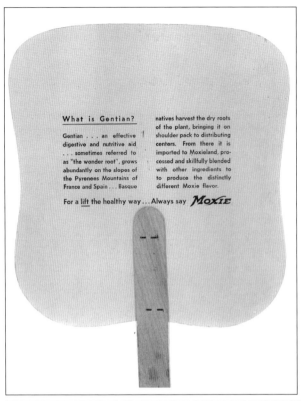

Eventually, the approach of using secretive ingredients and mysterious formulations for soft drink beverages gave way to increased transparency and open disclosure. The back side of this c. 1940s Moxie Boy cardboard fan—possibly the last one the company produced—proudly informs the public of the digestive benefits of the "wonder root" called gentian (left). Gentian morphed into an intriguing story to tell the public—hand harvested by Basque natives on the slopes of the Pyrenees Mountains. A more critical supply issue faced by Moxie and other soft drink bottlers was the shortage of sugar experienced during World War II. The public was acutely aware of wartime difficulties experienced by manufacturers and processors, as it was issued ration books for a number of commodities, including sugar. (Below, courtesy of the Library of Congress.)

In 1923, the Moxie Company under Frank M. Archer applied to the US Patent Office for trademark protection, noting its successful 1905 trademark registration and its use of the Moxie name dating back to 1885 (right). It is interesting that in legal filings such as this one, the Moxie Company did not use 1884 as its date of origin, as claimed in many advertising pieces. For most of its first five decades of existence, Moxie never experienced a shortage of imitators trying to capitalize on its name and extensive brand recognition with the consuming public. The company was constantly vigilant in protecting its brand by identifying trademark violators and bringing them to court. A succinct warning message using the pointing Moxie Boy graced the interior pages of the published booklet *This Book About Substitution Law* (below).

Registered Sept. 9, 1924.   TRADE-MARK 189,066

UNITED STATES PATENT OFFICE.

THE MOXIE COMPANY, OF PORTLAND, MAINE, AND BOSTON, MASSACHUSETTS

ACT OF FEBRUARY 20, 1905.

Application filed March 26, 1923. Serial No. 178,054.

MOXIE

STATEMENT.

To the Commissioner of Patents:

The Moxie Company, a corporation duly organized under the laws of the State of Maine, and located in the City of Portland, county of Cumberland, in said State, and doing business at 61-71 Haverhill Street, Boston, Massachusetts, has adopted and used the trade-mark shown in the accompanying drawing, for a NONALCOHOLIC, MALTLESS CARBONATED BEVERAGE AND SIRUP FOR MAKING THE SAME, in Class 45, Beverages, nonalcoholic, and presents herewith five specimens showing the trade-mark as actually used by applicant upon the goods, and requests that the same be registered in the United States Patent Office in accordance with the act of February 20, 1905, as amended.

The trade-mark has been continuously used and applied to said goods by the applicant and its predecessors from whom title was derived since the year 1885 and such use has been exclusive. The applicant is the owner of trade-mark registrations No. 12,565 to A. Thompson and No. 62,295 to the Moxie Nerve Food Company of New England.

The trade-mark is usually applied or affixed to the goods by placing thereon a printed label on which the trade-mark is shown.

Applicant hereby appoints Oliver Mitchell, Everett D. Chadwick and Everett E. Kent, constituting the firm of Mitchell, Chadwick & Kent, (Reg. No. 8481), of 99 State Street, Boston, Massachusetts, its attorneys, to prosecute this application for registration with full powers of substitution and revocation, to make alterations and amendments therein, to receive the certificate of registration and to transact all business in the Patent Office connected therewith.

THE MOXIE COMPANY,

By FRANK M. ARCHER

Vice-President.

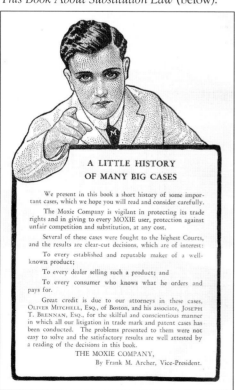

A Typical Instance of a Customer's Vigilance and the Result

MISS SHAW'S LETTER

Cathance Lake, Cooper, Maine.
August 13, 1917.

The Moxie Company,

Dear Sirs:

On August seventh I bought three bottles of Moxie at Harry Lombard's store, in Meddybemps, Maine. When I got them back to the camp I noticed they were *without* labels and each bottle had the same sort of cap to it. (I am sending you one of the caps.) One of the bottles was without doubt a regulation Moxie bottle, stamped with your mark, but the other two were marked "Four Crown Soda Water, Clark's Harbor, N. S., M. A. Nickerson." They all contained the same kind of drink (imitation Moxie) which made two members of the party extremely ill for about six hours.

We have been drinking Moxie all our lives and it has never before made us ill.

I do not wish to make any claims but I do hope that you will follow this up, for the vile stuff was bottled in one of your bottles which I will be glad to send to you upon request. My reason for writing this is to save someone else a similar experience.

Very sincerely yours,

S. F. SHAW.

The Moxie Company booklet, *This Book About Substitution Law,* presents a number of trademark infringement cases and their successful prosecution of violators. In Volume III from the late 1920s, a letter from a loyal Moxie drinker in rural northeast Maine exhibited above-and-beyond detective work and reporting of a local imitator's "vile stuff" (left). Such a degree of customer loyalty for Moxie was, and still is, quite common. Moxie look-alikes included the Lynn (Massachusetts) Bottling Works' Universal Appetizer and the Witch City Appetizer of Salem, Massachusetts, both employing orange labels with black lettering (below). Courts found in favor of the Moxie Company in both instances.

These bottles found by the Supreme Judicial Court of Massachusetts to be imitations of MOXIE bottles, and publicly destroyed, by order of the Court.

A number of Moxie imitators were fairly blatant in naming their beverage, presumably to either associate themselves with a legendary brand or simply to confuse an unwary consumer into purchasing their product. The Visner Bottling Company's Nox-All Nerve Food with Moxie-looking quart bottles and paper labels was one of the boldest competitors (right). The Massachusetts courts ordered Visner to cease producing its beverage and to deliver all existing bottles and labels for subsequent destruction. Noxie-Kola was produced by a Canadian bottler with claims to relieve "sleeplessness, nervousness and mental exhaustion" as well as improving the appetite and aiding digestion (below). In addition to metal signs, Noxie-Kola's advertising included horse-drawn bottle wagons.

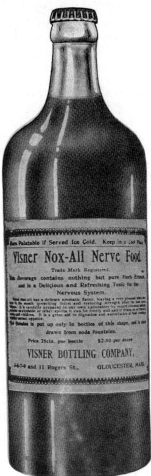

Commonwealth of Massachusetts.
SUPERIOR COURT
THE MOXIE COMPANY *v.* VISNER BOTTLING COMPANY.

Defendant enjoined from selling a beverage called "Noxall Nerve Food" or any other beverage as and for "MOXIE" or from suggesting or facilitating its substitution by retail dealers; from selling any beverage under the name or mark "Noxall Nerve Food."

Ordered, that the defendant deliver up for destruction counterfeit bottles and labels used in marketing "Noxall Nerve Food."

OLIVER MITCHELL,
    Attorney for Moxie Co.

J. S. SILVERMAN,
    Attorney for defendant.

No imitator gave Moxie more trouble than the Providence, Rhode Island-based brand called Modox. Modox was started in the early 20th century by James Stephen Barry and hailed itself as "The New Nerve Drink," with the claims it was made with Indian herbs. The brand often used the images of warrior chiefs in its advertising. In this image (left), the fictitious Indian Chief Modox was prominent, perhaps as somewhat of a counter play to the Moxie Company's Lieutenant Moxie equally fictitious icon. The Moxie Company took Modox to court and, on May 20, 1907, secured a knockout judgment against its competitor.

Modox was first bottled in the Nickel-Tone manufacturing plant at 17-19 Mathewson Street in Providence. Sometime later, the operation was moved to the site of the Hanley Brewing Company on Fountain Street in Providence. Despite the courts siding with the Moxie Company in their trademark infringement suit, the Nickel-Tone glass bottles were distinctly different from Moxie's bottles. However, the courts made Modox purge their brand from commerce, and the embossed "MODOX" was physically ground off the Nickel-Tone bottles' shoulders (right). This wooden crate (below) with the "It Strengthens The Nerves" message was given a thin coat of red paint across the still-visible "MODOX" lettering. The crate's inside stenciling was untouched, however, and still proudly heralded "MODOX" in clear red letters.

Frank Morton Archer worked his way up the proverbial ladder at Moxie operations from an off-the-street office worker in 1896 to sales manager and on to the chairman of the board in 1930. In 1928, he was awarded 2,000 shares of stock as recognition of his stellar contributions to the company's successes (above). But in 1937, Archer passed away at his home in Brookline, Massachusetts, leaving the company devoid of its long-term visionary, strategist, and chief promoter. His son Frank Archer Jr. took the helm (below, center), here seen recognizing chief chemist James Penny (right) and brother Arthur Penny (left), Moxieland's plant superintendent in 1940. However, Frank Archer Jr. lacked the skill sets and drive of his father and sold the Moxie Company to American Dry. The future of the iconic Moxie brand was uncertain.

Four
The Marketing Machine

The Moxie Company was a relentless marketing machine, taking full advantage of all available media of the day. A fan sent the company this photograph of a wholesome young girl with a Moxie sign that proclaims, "The Standard Family Beverage." To her right stands a bottle cooler, popular for keeping drinks cool in shops and stores.

Moxie was not typically pitched as a children's beverage, but seen here (left) is a print advertisement extolling its health virtues for the youngsters by strengthening those nerves while still remaining a "thirst-quenching" drink. To reinforce the safe and healthy aspects of Moxie, the public is reminded that it is sold by "Grocers, Druggists, and Dealers in temperance beverages." On this cardboard fan (below), the Moxie Boy beckons consumers with his pointing finger while a young girl on his knee holds a full glass of the beverage. The reverse side of this fan displays children in a classroom setting with a bottle of Moxie, not an apple, visible on the teacher's desk.

With Frank Archer's creation of the new seven-ounce Kid Moxie single-serve bottle, the previous cute and stylish-looking children were replaced with a likable but somewhat streetwise youngster, the Moxie Kid (right). The print advertisement highlighted both the availability of the new carrying bag that could hold six of the small bottles and its recyclability, citing "a hundred uses later." In the other advertisement (below), the reader's attention is drawn to the three available bottle sizes: 7, 16, and 26 ounces. The friendly Moxie Dog makes an appearance accompanying the Moxie Kid. Both print advertisements employed the 1920s-era taglines of "If At All Particular," and "OK'd By Millions."

The first Moxie Boy likely appeared in 1907, as the Moxie Company was recasting its advertising campaigns after the 1906 passage of the Food and Drug Act (left). Wearing a black pillbox cap, with the older plain "X" logo, the Moxie Boy cutouts ranged from three feet high to smaller soda countertop sizes. The "Learn to Drink" message was perhaps an acknowledgment that as a refreshment, not a remedy, Moxie's taste needed to be acquired. The Moxie Boy's image also appeared on metal pins and die-cut cardboard wall hangings. The image in the postcard (below) displays a newly adopted logo on his cap with a more stylized script and elongated "M" and "X," still in use today. This version of the Moxie Boy image was especially popular between 1908 and 1911.

In 1911, the Moxie Boy image was considerably altered (right). He is now dressed as a soda fountain attendant in a white laboratory-style coat and a necktie with a clearly visible "M" displayed. The new Moxie Boy had an intense stare directly at the viewer while his right hand pointed toward you in a commanding fashion. Indeed, the admonition emanating from this serious figure was usually the directive, "Drink Moxie!" The new image quickly became a classic trademark for the Moxie Company and was used on a wide array of advertising, as seen here on a cardboard fan with the 100 percent assurance boldly stated (below).

Many believed that Frank Archer himself was the model for the clean, wholesome gaze of the new Moxie Boy introduced in 1911. Archer and others denied this, but the idea was partly fueled by his name appearing so frequently in company advertising, including on Moxie Boy pieces. He may have done little to diffuse such speculation since any publicity helped to sell more product. As the Moxie Boy grew in familiarity, people everywhere swore they knew whom the image was modeled after. In actuality, the image was most likely simply that of a model used by the American Lithographic Company in developing the advertising campaign for the Moxie Company. Many companies would go on to copy the pointing, staring Moxie Boy for use in their own advertising. The well-known American illustrator James Montgomery Flagg may have used Moxie's pointing man as inspiration for his 1916–1917 now-famous Uncle Sam recruiting poster, although Flagg once cited a British illustrator as the source of his idea. (Courtesy of the Library of Congress.)

Moxie Girls were a staple of the company's advertising from the founding years up to World War II. A wide variety of attractive women came to adorn a series of colorful die-cut cardboard fans from 1915 through the 1930s. Silent film actress Muriel Ostriche was the first Moxie Girl on a fan in May 1915 (right) and was a Frank Archer favorite, appearing on at least four other fan variations. Muriel also appeared on cardboard cutout displays and later graced a series of Moxie Girl dishes. Widely popular, she made appearances in Boston, Lowell, and Waltham, Massachusetts, and even Nashua, New Hampshire, for the Moxie Company. The fan's tagline, "Clean-Wholesome-Refreshing," was often used with Moxie Boy images as well. The unknown 1924 lass (below) was also used on cardboard cutouts and store displays of the time.

Ireland-born actress Eileen Percy did not get as much exposure with Moxie, as did Muriel Ostriche, but her film career was much more successful; she co-starred with Douglas Fairbanks at the age of just 18 years old. She had a total of 68 career appearances in films and was considered one of the best-dressed on the New York stage. In this 1923 Moxie image (left), Eileen sports Mary Pickford–type curls and wears a Moxie Boy necklace. The unknown 1925 Moxie Girl gazes longingly at an image of the Moxie Boy in her compact (below). Moxie cardboard fans were typically giveaways at fairs, parades, carnivals, and similar public gatherings. Surprisingly, many have survived in good condition in the hands of Moxie memorabilia collectors.

In addition to Moxie Girls with stage and screen reputations, the Moxie Company drew in many other recognized personalities and celebrities in its print advertising campaigns. Among these celebrity figures were veteran vaudevillian and actor Ed Wynn (right); and Billy B. Van (below), renowned for his work in vaudeville, burlesque, and the New York stage. Frank Archer was friends with a large contingent of these and other celebrities, such as Mary Pickford and George M. Cohan. When celebrities were in the Boston area, Archer would often entertain them in his home. In turn, his celebrity friends would mention Moxie from the stage or even work Moxie into their professional acts.

In 1904, the Moxie Company commissioned a song, "Just make it Moxie for Mine," and the sheet music was published just in time for the St. Louis World's Fair (left). This may have been the very first singing commercial for a soft-drink beverage. In 1921, the company copyrighted a different song, simply called "Moxie" and recorded and released a 10-inch, 78-rpm record, which reportedly became a popular dance tune in its time (below). In one week in 1926, the Moxie Company gave away some 2,000 copies as part of its promotional efforts. Lyrics included, "For Moxie has a flavor all its own, good and pure, safe and sure, let everyone proclaim its name and fame."

Elf-like Brownies were the work of illustrator Palmer Cox, who first created jovial bands of these mischievous woodland spirits in 1880, eventually publishing some 13 Brownie books. The Moxie Company's use of these little pranksters in print advertising was fairly brief, likely only between 1905 and 1908. A Brownie dressed formally in coat and tails points with pride to the always-recognizable Moxie bottle and label.

The Moxie Maid—and especially, the odd-looking Moxie Butler—were used starting in 1919 and up through the end of the 1920s. Modeled after stage comedian and friend of Frank Archer, Raymond "Hitchy Koo" Hitchcock, the Moxie Butler was patented in 1922. Life-size carved models of the Hitchy Koo figure were used in theater lobbies everywhere, and smaller versions were produced to give away to Moxie distributors as product promotions.

If the Moxie "Foxtail" logo is an accurate guide, this metal drink tip tray was likely produced and used sometime between 1904 and 1907 (left). When the Moxie Company was granted a new patent in 1907, the new logo with the tail on the first part of the "M" and on both ends of the "X" was adopted and is still used today. Note the National Health Beverage message on the tray rim. Made from Moxie extract, candy in rectangular tins was sold for 5¢ in 1931 (below). The reverse side claims, "Frank Archer says: A pure delicious confection with the same distinctly different bitter sweet flavor as Moxie the beverage."

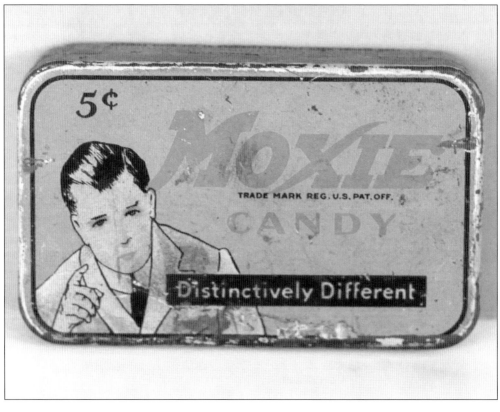

The Moxie Company often printed materials in various forms explaining its proud history. This three-panel booklet provided plenty of Moxie nostalgia inside but reserved the cover for a stylized illustration of the Moxie Bottle Wagon (right). The bottle-wagon concept was the creation of Francis E. Thompson, (eldest son of Augustin) and Freeman N. Young in July 1886, just one year after the "birth" of Moxie Nerve Food. The first wagons were simple four-wheel, horse-drawn carts with a large wooden bottle positioned on the rear where a uniformed attendant could sit and dispense glasses of the beverage for 5¢. A patent was secured for the bottle wagons in 1889 (below).

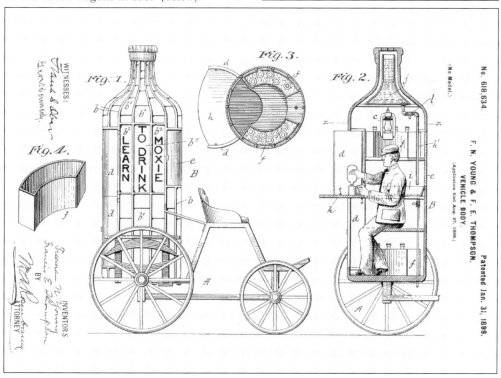

At least some of the Moxie bottle wagons were made by the Dole and Osgood Carriage Works of Peabody, Massachusetts, one of four carriage makers in that town. Three of Moxie's own are seen here in front of that carriage works, most likely around 1900 (above). Reportedly, the average cost to construct such a wagon was $500. Many variations of the bottle wagons were made over their three decades of popularity: a hand-pulled version, a pony-pulled version, and at least one tricycle mounted version. The small two-wheel bottle wagon seen in the urban street next to a constable (below) appears to be a hand-drawn version.

Some of the wagons the Moxie Company had constructed were fairly ornate, while others were more rustic. The smartly dressed girl in the riding seat, the formally attired gentlemen (perhaps a coachman) in the top hat alongside the carriage, and the patriotic bunting adorning the carriage wheels likely indicate the bottle wagon's participation in a parade.

In addition to plying city streets, the Moxie Company dispatched its wagons to travel from town to town and were frequently staged at fairs, parades, and similar community events; documented Northeast destinations included Coney Island, Atlantic City, and Washington, DC. They could even be rented out for private parties or for company outings. Note the image of the original Moxie Boy on the side of the bottle.

83

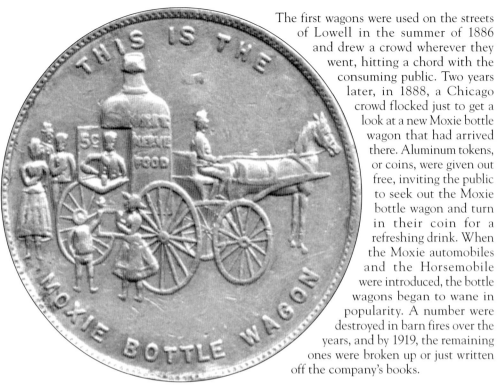

The first wagons were used on the streets of Lowell in the summer of 1886 and drew a crowd wherever they went, hitting a chord with the consuming public. Two years later, in 1888, a Chicago crowd flocked just to get a look at a new Moxie bottle wagon that had arrived there. Aluminum tokens, or coins, were given out free, inviting the public to seek out the Moxie bottle wagon and turn in their coin for a refreshing drink. When the Moxie automobiles and the Horsemobile were introduced, the bottle wagons began to wane in popularity. A number were destroyed in barn fires over the years, and by 1919, the remaining ones were broken up or just written off the company's books.

The Moxie Nerve Food Company purchased its first Stanley Steamer automobile in 1902 from the Stanley Dry Plate Company, then a Maine firm. Moxie was quick to adopt this new motorized technology and use it to its full advantage. For many small-town children, the Moxie automobile may have been the first "horseless carriage" they ever viewed. The Moxie delivery vehicle, photographed in 1906 (above), was part of a group, each sporting a gold-painted foxtail-style logo (only used 1904–1907). The Rambler vehicle (below) was part of a six-Rambler and one–Stanley Steam string parked in front of the Thomas B. Jeffery's New England Agency in Boston in February 1906. In the year prior, the Moxie Company dispatched 14 of its automobiles to caravan from Boston to New York City in a show of promotional force.

Moxie delivery trucks, like this c. 1908–1910 Buick (above) with its white paint scheme, convertible top, and "captain's railing," were a common sight in New England. In addition to bottled beverage, they also carried promotional items for the drivers to give out to retailers. Most vehicles also had a cutout Moxie figure adorning each running board. The New York City branch of the company utilized White Motor Company trucks for beverage deliveries (below, around 1914) with the customary cutouts gracing the vehicles' sides. Moxie vehicle drivers were typically referred to as the "Moxie Man." Company literature of the era claimed that in its first 10 years of employing 54 steam, electric, and gasoline vehicles, a quarter of a million store visits were made, and the total vehicle mileage amounted to the equivalent of 51 times around the earth.

Frank Archer is credited with the concept of mounting a horse on an automobile chassis with a driver sitting in the saddle and controlling the vehicle via a steering wheel. In the summer of 1916, the Moxie Horsemobile made its debut, and a patent followed in February 1917. The first horses were harness makers' plaster display models mounted on a vehicle frame. The plaster would crack with the vibrations from the road, and drivers carried white tape with them for making repairs. Later models included wooden horses from England and then molded aluminum bodies. The photograph (above) shows an unknown driver on his white horse mounted on an Essex automobile with a number of curious onlookers. With learned experience, the Horsemobile evolved in complexity, and the company's basic patent was renewed in 1936 (below).

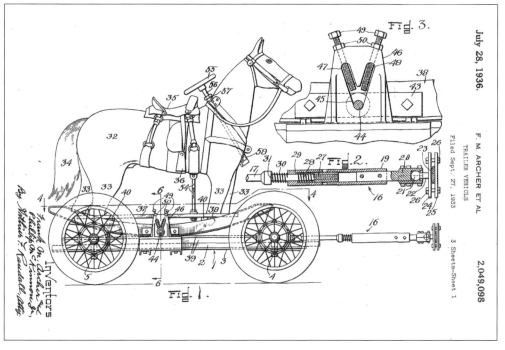

With their eye-catching appeal, the distinctly different Horsemobiles were an instant promotional phenomenon. A May 1919 story in the *Manchester Union* newspaper reported on a visit by one of the Moxie vehicles, "The Horsemobile attracted a great deal of attention, and it's entirely new to anything that has even been displayed in this City." A number of Horsemobiles were built over the years, and some remained in use up to the 1950s. In addition to the Essex, automotive chassis included Metz (constructed in Waltham, Massachusetts), Dort, Maxwell, Trumbell, Rambler, LaSalle, Buick, and Rolls Royce. In his promotional film for Moxieland, Frank Archer showcased a line of Horsemobiles matched to a Buick, a Rolls Royce, and four LaSalles (above). A Moxie brochure was proud to highlight the Army's use of Horsemobiles in recruiting efforts (below).

THE MOXIE HORSEMOBILE
Shown in use by the United States Army Recruiting Service

In 1919, the H.D. Beach Company from Ohio came out with a toy Moxie Horsemobile. The toy consisted of brightly lithographed thin steel panels that the buyer assembled by attaching the wheels by bending metal strips into their proper position. The toy Horsemobile in the photograph consisted of a white horse and driver on a bright blue automobile, possibly an Essex.

This attractive young cowboy and his rocking horse pictured on a cardboard fan from 1922 reflect the popularity of the Moxie Ponycycle during the heyday of the bottle wagons. It also gives a nod toward the promotional craze that Moxie set in motion in 1917 with the advent of the Horsemobile and uniformed driver.

The Moxie Company's promotional efforts often had a tie-in to popular sports. A newspaper article of July 18, 1887, refers to the Moxie Nerve Food Baseball Club and their victory the prior day over the Athletics by a score of 5 to 3, indicating a likely company sponsorship situation. The Moxie Company also placed an advertisement in the very first issue of *Baseball Magazine* in May 1908 that stated, "Athletes and Businessmen Drink Moxie in Order to Eat Better, Sleep Better, and Feel Better." At the turn of the century, the term "moxie" was beginning to be used to describe athletes with skill, energy, and enthusiasm. At least three major league ballplayers of the era adopted the nickname: Mark Garfield "Moxie" Manuel (1908 Chicago White Sox), Merton Merrill "Moxie" Meixell (1912 Cleveland Naps), and Edward George "Moxie" Divis (1916 Philadelphia Athletics). Unfortunately, none of the "Moxies" were keepers; Meixell and Divis each participated in only a single major league game. The name and location of the c. 1900 team in the photograph is unknown, but the sponsorship is clear.

From print advertising to store displays to bottle wagons and gasoline vehicles to billboards and painted roofs, the Moxie Company's sales staff sought out every opportunity to keep the Moxie name in the public's eye and sell product. The leaping baseball player saving the sure-thing home run (above) is using a Kid Moxie bottle display for support and reminding thirsty sports fans that Moxie equates to Nerve. A plug for the Pureoxia flavors is also worked into the promotional piece. One of the Moxie Company's popular sports giveaways was the cardboard scorecard disk with a chart on the reverse for keeping track of the baseball game's inning by inning scores (below).

The popularity of prizefighting, especially in the first half of the 20th century, was also not overlooked by the Moxie Company's marketing teams in developing their advertising themes (left). Here, the clear message is winners are winners because they have the "moxie" to become champions in their athletic endeavors. The Moxie Tennis Girl's image (below) was used extensively in the early 1940s on cardboard cutout displays for store windows, cardboard bottle hangers (shown here), and in newspaper advertisements. Consumers were sent the message that having "moxie" applied not only to brutish, hairy prizefighters, but also to attractive, athletic females.

The Moxie Company ran a series of print advertisements in many newspapers starting in the summer of 1926, pairing their beverage (especially in seven-ounce Kid Moxie size) with specific culinary suggestions. In this photograph (right), the viewer is encouraged to enjoy a Moxie along with a famed New England steamed clam dinner ("little necks" in some New England parlance). Continuing on with the seafood theme, Frank Archer's marketing machine suggested that nothing could go better with a platter of fish than having an ice-cold Moxie (below). Surely, some of these admonitions much have seemed a stretch to even ardent lovers of the Moxie beverage.

Nothing says classic Yankee-country food more so than the traditional New England boiled dinner, a staple of Sunday's formal sit-down meal with the extended family gathered around the table (above). The advertisement's tagline's claim of "They are fine together" seems a bit tenuous, as if trying to search for something reassuring to say to convince the newspaper's readership. Likewise, the pairing with hot or cold lamb could not have appealed to that large an audience (left). Other print advertisements of the day suggested enjoying Moxie with roast beef, duck, lobster, spaghetti, chop suey, hot dogs, ham and cheese, oysters, minced pie, ice cream, and that New England classic dessert, Indian pudding.

The Moxie Company began using Santa Claus in its Christmas season print advertising in the 1920s, probably about the same time as Coca-Cola did. Whether one of the beverage companies copied the other is uncertain, but the Santa Claus image was growing as an advertising tie-in. The most common image in use in the first decades of the 1900s patterned Santa's looks along the line of the Thomas Nast images—a little elfish. It would not be until the early 1930s when Coca-Cola developed the jolly, plump Santa image in its advertising that there was a clear standard-bearer of what Father Christmas should look like—at least in American culture. In this Moxie advertisement, a clever Santa sails over a typical New England village scene by commanding not a team of reindeer, but a Reindeermobile fashioned after one of the company's Horsemobiles. The former existed only in the minds and pens of the company's marketing team, however. Note the Moxie song sheet and a bottle of beverage in Santa's sack.

A rather dour-looking Santa Claus hands out glasses of Moxie to a line of children (left) while the advertisement's printed message humorously employs parents to let them indulge. After all, the beverage is "scientifically blended from nature's own ingredients." The Moxie Boy peers out from behind a Santa Claus mask to wish the readers a Merry Christmas (below). Unlike most of the company's Christmas season advertising, this one presents a collage of well-recognized Moxie advertising symbols but no seasonal tie-in other than the greeting itself. The message, as in most of Moxie's Christmas season print advertisements, suggests ordering a case for the coming holiday, including serving the drink at Christmas dinner.

In addition to actively plying state fairs, parades, and other leisure-time gathering places, one of the Moxie Company's sales strategies was to associate itself with recreation of all types. Moxie became thoroughly visible at the seashore, mountains, amusement parks, picnic areas, and wherever the public gathered for recreation. This c. 1910 postcard image of the square in seaside Houghs Neck, a section of Quincy, Massachusetts, shows Moxie signage on two separate establishments (above). Another postcard of Houghs Neck from the same timeframe (not seen here) shows a third tourist-related building with a prominent rooftop Moxie sign. The refreshment stand built out over a marsh (below) seems to be messaging that it sells Moxie tonic exclusively. With the use of the word "tonic," one can be assured this is establishment is located somewhere in New England.

Some of the Moxie Company's print pieces that were oriented toward the recreation-minded would suggest shipping a "good supply" of Moxie to one's summer home so it would be waiting there upon their arrival. At least one advertisement seemed to play on urban-dwellers fears of being away from the security of the city, claiming that drinking Moxie would eliminate the risk from "impurities in doubtful drinking water." In one advertisement, having Moxie meant one was braced up enough to dive headlong into the summer swimming hole (left). In this c. 1920s summer scene at New York's Coney Island, the Moxie refreshment stand is well positioned directly across from the immensely popular Feltmans Restaurant, where Charles Feltman created America's first hot dog (below).

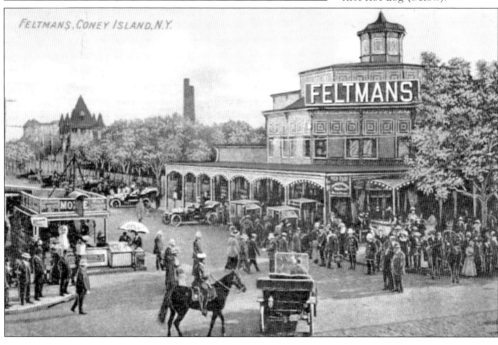

32 foot Moxie Bottle, Tall as a Three Story House, Pine Island Park,
Published by Graves & Ramsdell   Manchester N. H.

The Moxie Company always thought big and bold when it came to promotions. Bottle wagons were popular, so why not go even bigger? In 1907, the company built the first of two giant wooden replica beverage bottles. The first bottle stood 32 feet high and was built as a refreshment stand from which an attendant could dispense glasses of cold Moxie to the thirsty public. This bottle went on display at a Boston Food Fair, was used subsequently at Coney Island, and then moved to Pine Island (amusement) Park in Manchester, New Hampshire, in 1910. The bottle remained in service until 1919. Chapter 6 details the later history of this unique advertising structure. A larger bottle constructed in 1911 was billed as "the largest bottle in the world" and exhibited at the Domestic Science and Pure Food Exposition in Madison Square Garden. It is believed to have been displayed at Savin Rock Park in Connecticut in 1915, but the final fate of that bottle is unknown.

Only infrequently did the Moxie Company steer its advertising dollars toward print pieces or promotional items that projected a theme of romance. Maybe the dyed-in-the-wool Yankees who ran the enterprise through all those decades learned that their customers responded more favorably to other themes. In this late-1800s trade card (left), Moxie seems to be only one of the interests shared by the attractive and impeccably dressed couple. Note the older Moxie logo on this card. Interestingly, Moxie did not commonly use trade cards, strongly preferring other forms of print advertising. On the reverse of one of the Muriel Ostriche cardboard fans is an idyllic country club courtship scene complete with a romantic Moxie canoe ride and beverages (below).

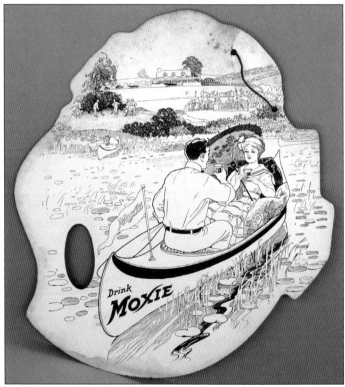

Moxie Girls were almost always attractive, sporting wide smiles and presented in casual settings. That makes this advertisement of a smartly dressed, no-nonsense female office worker most unusual (right). This Moxie Girl from the early 1940s is a bit more Moxie conventional in appearance (below). The attractive solo canoeist sports a modern two-piece bathing suit (only in fashionable use since the mid-1930s) and is being pursued by two athletic hunks in a second canoe who exclaim, "She's Got Moxie." The "Frank Archer says" line actually refers to Frank Archer Jr., who took over the Moxie Company after the passing of his father in 1937. The six-pack carton displayed in the advertisement was a Frank Archer Sr. innovation in the 1930s and eventually became a beverage industry standard.

The girl with Moxie is the girl who gets ahead. If you want to be popular, get the pep and sparkle that have made Moxie popular. Get Moxie—today.

FRANK ARCHER SAYS: Moxie leaves a clean taste in your mouth. Try it and see how pleasant and refreshing it is compared with sweeter beverages.

From its inception and beyond, the Moxie Company always reflected a sense of national pride and outright patriotism. Such messaging carried into World War II, and company advertising often presented attractive, confident females serving in various military roles. A WAVE (Women Accepted for Volunteer Emergency Service) knows to say "Moxie" for nerve as a gawking sailor affirms her spunk in correcting a Navy admiral (below). The civilian population (left) during this period were asking how they could contribute to the war effort, so a Civil Defense officer suggests an answer using a patented company proclamation of the times, "What this country needs is plenty of Moxie."

"A great Army favorite" is the message from this Army WAC (Women's Army Corps) from behind the wheel of her jeep (right). Interestingly, Moxie makes a point of a "not-too-sweet" drink with a "dry flavor," a counter to the sentiment by many that Moxie's taste had to be acquired. Note the admonition by Frank Archer that war metals were in short supply, so purchasing large bottles of Moxie saved resources. The hardworking male civilian worker population was also acknowledged in the Moxie Company's wartime advertising (below). Moxie could put "pep in your step," making one more productive. Such print pieces often encouraged the public to "Buy War Bonds" in support of the war efforts.

The Moxie Company participated in numerous wartime efforts, including scrap metal collections. In a promotional shot, Frank Archer Jr. (standing far right) contributes to a metal-recycling campaign with these 1941 license plates. Note the Moxie truck staged in the background. The World War II era brought numerous problems for the Moxie Company. The death of Frank Archer Sr. in 1937 left a tremendous creative hole in the enterprise that his son was unable to fill. Archer Jr. openly expressed concerns for the future of the brand, and he left the company in 1943. Wartime sugar rationing also hurt the company with quality and taste sometimes becoming problematic. The public was accustomed to wartime disruptions and sacrifices, but Moxie was losing ground to other brands, most notably Coca-Cola. Long passed was the glory era when Moxie actually outsold its rival Coke. A rescue plan and new vision were sorely needed.

Five

STRUGGLES AND NEW OWNERS

Frank Archer's nephew Orville Purdy joined the Moxie Company in 1922 at the age of 17 and rose through the ranks of sales manager to become vice president. Many claimed he kept the company ship afloat during some of its roughest years. In 1940, Purdy gave the iconic Moxie Boy a new look with this Frank Sinatra–like image, reportedly to springboard on the young crooner's growing popularity.

Without solid industry leadership, Moxie management made several questionable moves. Frank Archer Jr. (above, seated far right) sold assets to American Distillers, somewhat shoring up the company financially. In 1948, New Moxie was introduced in an attempt to target younger consumers whom management felt wanted a sweeter, less bitter flavor to their soft drinks (right). The change alienated traditional Moxie drinkers and completely backfired on the company. Rising costs and diminishing profits forced Moxie to abandon its once heralded Moxieland plant in favor of a number of smaller facilities in the Boston area, which now produced concentrate and syrup for franchised bottlers. The Pureoxia brand was sold to a local bottler who subsequently dropped the line entirely in the early 1950s.

In 1953, the area of Jamaica Plain, where Moxieland was located, was taken over by the city to make room for an urban development project. This was a mixed blessing, as it provided the Moxie Company with much-needed capital to counteract diminishing profits. Management acquired a modest 4,000-square-foot building in suburban Needham Heights adjacent to the high-technology-inspired Massachusetts Route 128 (now Interstate 95). The building was topped with a large Moxie sign and a huge replica bottle next to the building, both quite visible from the highway in front. The two-level structure at 295 Reservoir Street was now known as the Moxie Office and Laboratory. The building housed office and laboratory staff for the production of concentrate to distribute to their faithful cadre of franchised independent bottlers. The Moxie Company ceased all in-house bottling operations as it tried to regroup in its new location.

By the mid-1950s, the Moxie Company's once-proud fleet of Horsemobiles had dwindled down to just three vehicles, all LaSalles. One was used on occasion for promotional events in Boston and Maine coastal towns, often driven or overseen by one of the company's managers, John Gillespie (pictured at left as a district sales manager in Springfield, Massachusetts). The horses were removed from the other two vehicles and sold, with one placed atop a used car dealership in Boston and the other going to a Cape Cod restaurant. Seeking other opportunities for promotion, in the late 1950s, Boston Red Sox's "splendid splinter" Ted Williams, a New England icon himself, was hired on to be a Moxie promoter, and an entire advertising campaign followed (below).

Ted Williams spent all 19 years of his major-league career as a left fielder for Boston, was a 19-time All-Star player, and had 521 home runs in his career. Williams's widely popular image was incorporated onto bottle labels, and many Moxie promotional devices and signs. In 1958, Moxie decided to further exploit Teddy Ballgame's popularity by introducing Ted's Root Beer to his fandom, even creating a painted wall in Boston's Fenway Park emblazoned with an advertising panel that survives to this day (above, with Merrill Lewis in "uniform"). Advertisements featuring Williams often promoted Moxie's healthful qualities (right). Ted Williams's relationship with Moxie ended in the early 1960s but was briefly renewed in the 1970s during the company's later Monarch era.

In 1958, the satirical MAD magazine started inserting the word "Moxie" into its pages. Then, in January 1959, the editors created a girlfriend for the iconic character, Alfred E. Neuman. When readers subsequently demanded a name for the girlfriend, editors responded with "Moxie Cowznofski." Sales of Moxie rose by 10 percent, pleasing the company, of course. In a quite-telling move, MAD editors responded to letters to the editor questions with, "Actually, we've been plastering "MOXIE" all over the magazine because we feel it ought to be restored to the popularity and place of prominence it once enjoyed upon the American scene. Yessiree, Let's 'BRING BACK MOXIE!' –Ed." The Moxie Company took advantage of this attention and adopted the "Mad about Moxie" tagline in its advertising, here on a label for No Deposit-No Return bottles (below). (Left, collage created by Jim Jansson.)

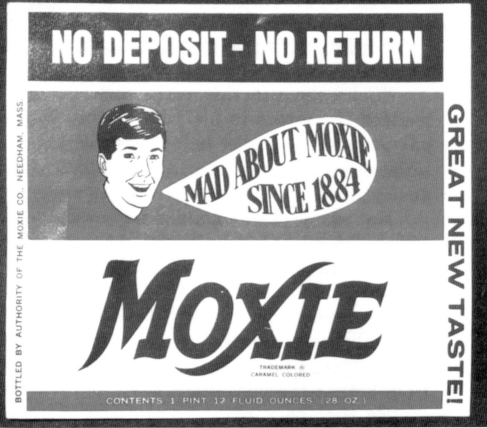

The Moxie Company's financial struggles continued with mergers and suitors considered and a chaotic "strategy" that included potato chips and chewing gum. In 1965, the Monarch Beverage Company was founded in Atlanta, Georgia, by Frank Armstrong, an advertising executive with soft drink–industry experience. Monarch's intended focus was otherwise-overlooked regional brands (NuGrape, Sun Crest, and Dad's Root Beer, for example), and Armstrong acquired the flailing Moxie brand in 1968. He subsequently moved operations to the Atlanta area, well away from the beverage's New England roots. Armstrong (above, far left) poses with several unidentified Monarch executives behind some of the company's branded soft drinks. In 1971, the company restored the last remaining Horsemobile to running condition—a LaSalle (below). Many traditional Moxie fans, however, recoiled at the horse's brown paint scheme, as the Moxie Company's steeds had almost always been white.

As it had in the late 1940s, Moxie once again tried a new sweeter formulation to attract a younger consumer, reflected in this "hip" 1968 advertisement displaying new dimpled bottles and "A Great New Taste" tagline (above). As was the case with the earlier venture, the new formulation and its promotional efforts were a dud. Traditional Moxie drinkers missed the original flavor and strongly rebelled (a lesson Coca-Cola had to learn with "New Coke" almost 20 years later). Monarch wandered far and wide during its ownership of the Moxie brand as it tried to expand market share and reach different consumers. For a time, a German beverage conglomerate, Eckes, even held a major interest in the brand moving the headquarters to California. A line of flavored seltzer water bottled in Louisiana and offbeat Moxie flavors were tried without great success (below).

During this period of attempted diversification, Monarch even ventured into the cereal marketplace. They produced a Tropical Fruit Granola Cereal on the West Coast under the Moxie label, complete with the pointing Moxie Boy and the tagline "Smart Foods for Smart People" (right). Wisely, Monarch reverted to "Old Fashioned Moxie" with a 1950s looking label and a somewhat more stylized Moxie Boy image (below.) Traditional Moxie continued to be sold primarily in New England during the Monarch ownership years and through contract bottlers. In 1985, the *Boston Globe* newspaper reported that the Moxie concentrate (called flavor number 012033 by the company) was shipped in five-gallon plastic pails to seven different contract bottlers in New England, who produced a total of half a million cases per year for consumers.

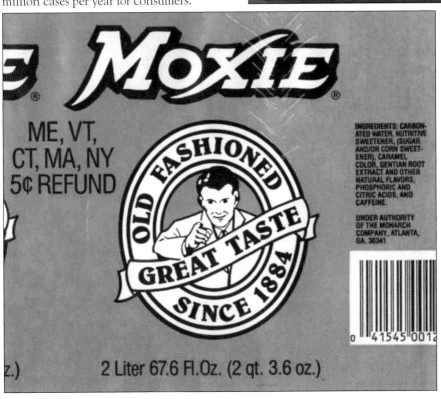

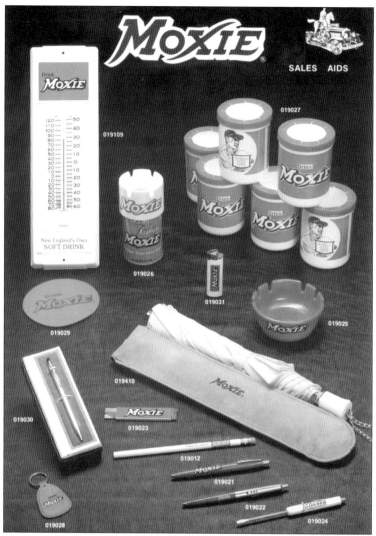

Through the 1980s and into the next few decades, Monarch maintained a low-key approach to promoting its Moxie brand and appeared to take its New England base for granted. In-store promotions were common, but expansive advertising budgets were not in Monarch's strategy. Sales aides did reflect a variety of giveaways destined for the hands of Moxie's contract bottlers to use for their promotional purposes but not for public sales (above). In the early 2000s, Monarch began to concentrate on its international markets, deciding to divest its domestic carbonated beverage products. The Moxie brand was put up for sale. Negotiations began with several potential buyers who hoped the brand could be reinvigorated under new ownership. The row of metal cans silently but symbolically charts the post-1950 Moxie journey from the pre-Monarch era up to today (as seen in the second photograph, from left to right). The story continues.

Six
A Distinctly Different Following

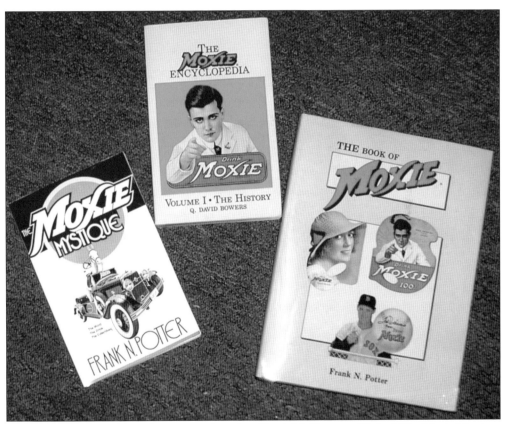

In 1981, just prior to Moxie's centennial, author Frank N. Potter published The Moxie Mystique (left), which started an unexpectedly large and sustained fan buzz. In 1985, Q. David Bowers published The Moxie Encyclopedia, Volume 1 (no volume 2) and Frank Potter released his second publication, The Book of Moxie, in 1987 (right). These books greatly helped generate a renewed interest in Moxie, especially in New England.

In 1982, Frank Potter held a book signing for his *Mystique* book in front of Frank Anicetti's Kennebec Fruit Company store in downtown Lisbon Falls, Maine. Anicetti's store sold Moxie for over 100 years and was known as the "House of Moxie." The book signing drew unexpectedly long lines of Moxie fans and was repeated in 1983, complete with reporters and TV cameras. This success launched the planning for 1984 celebrations in Lisbon Falls and Union, Maine (Augustin Thompson's birthplace). In early advertising, Moxie touted its patent year of 1885 but at some point, in the 1930s, began to use "Since 1884" (perhaps to best Dr. Pepper, which also claimed an 1885 origin). In any regard, 1984 was chosen to be Moxie's centennial year. In 1985, Moxie festivals were held in Lisbon Falls, followed by a "Moxie Day" at Clark's Trading Post in Lincoln, New Hampshire. Ardent Moxie fans Ed Clark and David Bowers had purchased the last remaining original Horsemobile from Monarch Beverages and brought it to Clark's tourist attraction in the White Mountains to join his extensive Moxie collection.

The former Lisbon Falls Frontier Days parade and events had morphed into the annual Moxie Festival, which has been the focus for all things Moxie ever since and may well be credited with refocusing the consumer public's attention back onto the Moxie brand. This event and its parade (above) have featured antique Moxie vehicles and associated memorabilia. Upwards of 40,000 fans attend annually during the second weekend of July each year. This bright orange-painted Volkswagen Beetle, the "Moxie Bug" (below) was owned by New England Moxie Congress (NEMC) member Russ Bilodeau for a number of years and made appearances in the annual Lisbon Falls Moxie Festival parades and other venues. The New Hampshire vanity license plate that Bilodeau held at the time is now registered to NEMC member and author Dennis Sasseville.

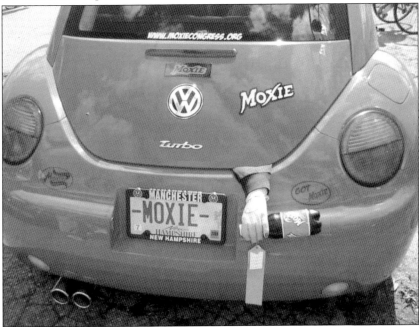

During the week prior to the Moxie Festival in Lisbon Falls, other Maine establishments typically host Moxie events, notably Moody's Diner in Waldoboro and S. Fernald's Country Store in Damariscotta (above). Frank Anicetti has been called "Mr. Moxie" by a generation of the beverage's followers who have been entertained by his Moxie-promotional speeches and enjoyed his homemade Moxie-flavored ice cream. Anicetti always found inventive ways to spread the Moxie spirit through his store's products and his personal brand of ambassadorship (below.) Sadly, Mr. Moxie passed away in 2017, but his now beautifully renovated former store is called Frank's, in his honor by new owners Traci and Tony Austin, who transformed it into an attractive pub with a Moxie theme.

In 1991, a loosely knit band of Moxie zealots and fellow travelers who collect Moxie memorabilia, promote the beverage, and enjoy parades and clambakes formed the NEMC, with its logo flying the foxtail version of their favorite beverage (above). High-energy NEMC members help organize the annual Moxie Festival in Lisbon Falls, Maine, as well as numerous other events throughout the northeast (moxiecongress.org). The Seashore Trolley Museum in Kennebunkport, Maine, was the site of the first meeting of NEMC, and its annual business meeting has been held there ever since on the Sunday after the Lisbon Falls event. NEMC president Merrill Lewis stands next to a 12-foot-high replica of an early Moxie bottle while the late Wil Markey sits astride a replica Horsemobile, one of several he built (below).

Honoring Moxie creator and native son Dr. Augustin Thompson, as well as recognizing Maine residents' traditional fondness for the beverage, Moxie was proclaimed as the official soft drink of the State of Maine. Gov. John Baldacci signed a proclamation into law on May 10, 2005, in the presence of the bill's supporters and Moxie lovers. Surrounding the governor are (from left to right) Gary Crocker, Maine humorist; Fred Goldrup ("Taurus the Clown"), the main proponent of the bill; State Representative Peggy Rotundo, the bill's sponsor; Frank Anicetti (partially hidden—"Mr. Moxie" of Lisbon Falls); state representative Chris Barton; Judy Gross (partially hidden); Sue Conroy, Moxie Festival chair; state representative George Bishop, Jr.; George Gross (partially hidden), president of Matthews Museum; and author Merrill Lewis, president of the New England Moxie Congress. Lewis reported that after the bill's signing, Governor Baldacci took a sip of Moxie, then rubbed his follicle-challenged head and exclaimed, "I think it's starting to work!"

The Moxie Bottle refreshment stand that once stood at Pine Island Park in Manchester, New Hampshire (page 99), was 32 feet high, 10 feet in diameter, and made of wooden sections that were pin-connected for dismantling and reassembly. In 1919, the structure was acquired by a private party and moved across the ice to become a unique part of a small summer cottage. A Boston newspaper gleefully carried the story of the bottle house at the time. The bottle survived the 1938 hurricane, and the silver cap served as a navigation aid to World War II aviators at nearby Grenier Army Airbase (now Manchester Airport). The structure was largely neglected by the 1980s (right) when along came NEMC in 1999 with a "Friends of the Bottle House" vision of saving and restoring the iconic bottle. NEMC President Merrill Lewis faces the camera (below).

Located in the birthplace village of Moxie creator Augustin Thompson, the Matthews Museum of Maine Heritage was identified as a logical home for the restored Moxie Bottle. First, however, substantial funds were required to restore the bottle itself, transport it to Union, and create a space for it to be displayed upright. NEMC members and other supporters worked tirelessly at fundraising and restoration efforts over a 10-year period. Supporters are pictured constructing the roof for the Moxie exhibit section of the small museum (above). Volunteer efforts culminated in the restored Moxie Bottle's placement at the Matthews Museum in 2009, where it stands surrounded by displays of Moxie memorabilia (left). Most of the memorabilia items were donated by loyal fans who appreciate and respect Moxie's unique place in the history of Americana.

Moxie continued to struggle as a regional brand owned and managed by Monarch Beverage's distant headquarters. In 2007, an unlikely savior appeared and purchased the Moxie trademark outright from Monarch and returned the brand to its New England roots. The rescuer was Cornucopia Beverages, a group owned by the Coca-Cola Bottling Company of Northern New England, Inc. (CCNNE), who had been bottling Moxie under contract to Monarch for several years. Renamed the Moxie Beverage Company and headquartered in Bedford, New Hampshire, its products are bottled at CCNNE's nearby Londonderry facility (above). Two other authorized entities supply Moxie in popular 12-ounce glass bottles for a nationwide market: distributor Orca Beverage Inc. of Mukilteo, Washington (below, left bottle and four-pack) and the franchise bottler Catawissa Bottling Company of Catawissa, Pennsylvania, (below, center bottle) alongside Moxie Beverage's bottled version (right).

A new slogan, "Live Your Life with Moxie," was introduced in advertising material, including the traveling billboard displays featured on the company's delivery trailers. Moxie has retained the essence of its advertising legacy by restoring the iconic 1911 Moxie Boy image and the traditional "Distinctively Different" slogan to their former prominence.

Over the past decade, the Moxie Beverage Company has produced many unique and interesting sales aids following the soft drink's long-established tradition. Aids and giveaways have included fans, baseball bats, bottle openers, and stickers distributed at the annual Moxie Festival and other venues. To the delight of fans, Moxie advertisements have also been aired during Boston Red Sox radio broadcasts.

The Moxie Beverage Company reestablished Moxie's patriotic legacy by supporting our valiant troops departing to and returning from combat areas. Working with members of the Pease Greeters, who meet each military flight passing through Portsmouth New Hampshire, over 10,000 cans of Moxie have been donated and distributed to service men and women (above) in recognition that America's brave troops both have—and deserve—a lot of Moxie! Modern graphics and social media are now used to promote Moxie to a new and growing audience as well as to its traditional customer base (below). The Coca-Cola Company of Atlanta acquired the brand in late 2018 and is expected to continue the beverage's cherished mystique. Moxie is America's oldest bottled soft drink, and it promises to be the "Distinctively Different" beverage of choice for generations to come.

BIBLIOGRAPHY

Anonymous. *This Book About Substitution Law, Volume III*. Boston, MA: The Moxie Company of America, 1929.

Baumer, Jim. *Moxie: Maine in a Bottle*. Camden, ME: Down East Books, 2012.

Bowers, Q. David. *The Moxie Encyclopedia, Volume I: The History*. Vestal, NY: The Vestal Press, LTD., 1985.

Clayton, John, "It takes Moxie." *In The City*. Portsmouth, NH: Peter E. Randall Publisher, 1993: 86–89.

Jansson, James A. *Moxie Time Line*. Shelton, CT: Bench Press, 2012.

Leheney, John, *To Maine with Moxie*. Independence, MO: LEW Printing, 2010.

Lewis, Merrill A. *Help Save the Moxie Bottle House*. Manchester, NH: Self Published, 2002.

———. *The Moxie Boy Centennial*. Manchester, NH: Published for New England Moxie Congress, 2011.

Paul, Jay. "Maine's Moxie Centennial." *Down East Magazine* (July 1984): 89.

Potter, Frank N. *The Book of Moxie*. Paducah, KY: Collector Books, 1987.

——— and Hank Stein. *The Moxie Mystique*. Brookfield, MO: Donning Company/Moxibooks, 1981.

———. *Orville Purdy Talks About Moxie: As Told to Frank N. Potter*. Paducah, KY: Self published, 1999.

Veilleux, Joseph A. *Moxie, an Acquired Taste*. Bloomington, IN: 1stBooks, 2003.

INDEX

Archer, Frank M., 30, 41, 46, 53–57, 60, 61, 63, 68, 71, 74, 75, 77, 79, 80, 87, 88, 93, 101, 103–105
Archer, Frank Jr., 68, 101, 104, 106
Armstrong, Frank, 111
Boston, 2, 11, 12, 30, 45, 49, 53–55, 58, 60, 75, 77, 85, 106, 108, 109, 113, 121
Coolidge, Calvin, 50
homeopathy, 11
Horsemobile, 84, 87–89, 95, 108, 110, 116, 119
Lisbon Falls, 116–120
Lowell, 12–14, 20, 25, 28, 30, 75, 84
Lyndeborough, 21
Maine, 9, 10, 18, 19, 27, 33, 41, 46, 48, 52, 64, 85, 88, 108, 116, 118, 120, 122
Matthews Museum, 120, 122
Moxie Bottle House, 99, 121, 122
Moxieland, 53–57, 61, 68, 88, 106, 107
Needham Heights, 107
Purdy, Orville, 105
Pure Food and Drug Act, 38–40
Pureoxia, 57–61, 91, 106
Roosevelt, Teddy, 31
Thompson, Augustin, 9–18, 20–28, 30, 33, 35, 51, 116, 120, 122
Williams, Ted, 47, 108, 109

Discover Thousands of Local History Books
Featuring Millions of Vintage Images

Arcadia Publishing, the leading local history publisher in the United States, is committed to making history accessible and meaningful through publishing books that celebrate and preserve the heritage of America's people and places.

Find more books like this at
www.arcadiapublishing.com

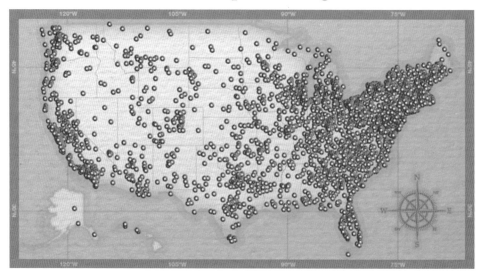

Search for your hometown history, your old stomping grounds, and even your favorite sports team.

Consistent with our mission to preserve history on a local level, this book was printed in South Carolina on American-made paper and manufactured entirely in the United States. Products carrying the accredited Forest Stewardship Council (FSC) label are printed on 100 percent FSC-certified paper.